Medical Radiation Biology

Medical Radiation Biology

Donald J. Pizzarello, Ph.D.

Professor of Radiology,
New York University Medical Center
New York, New York

Richard L. Witcofski, Ph.D.

Professor of Radiology,
Associate in Neurology
The Bowman Gray School of Medicine
of Wake Forest University
Winston-Salem, North Carolina

Second Edition

Lea & Febiger   Philadelphia

Lea & Febiger
600 Washington Square
Philadelphia, PA 19106
U.S.A.

Library of Congress Cataloging in Publication Data

Pizzarello, Donald J.
  Medical radiation biology.
  Bibliography: p.
  Includes index.
  1. Radiation—Physiological effect.   2. Radiation
—Toxicology.   3. Radiobiology.   I. Witcofski,
Richard L.   II. Title.   [DNLM: 1. Radiobiology.
WN 610 P695m]
QP82.2.R3P59   1982   612'.01448   81-19311
ISBN 0-8121-0834-5   AACR2

First Edition, 1972

Print No.   3   2

To our wives,
Nina and Jane,
with love

Foreword

The striking success of the first edition of *Medical Radiation Biology* is a tribute to its clarity and the fact that it filled a major need of new trainees in radiotherapy, radiodiagnosis, and nuclear medicine. The rapid accumulation of new material and the development of new concepts call for this second edition. Like the first, it will supply the new trainee with the essentials of radiobiology necessary for a solid foundation of his specialty. With this foundation, he will be better able to understand clinical concepts necessary for optimum patient care.

The ability of Drs. Pizzarello and Witcofski to appreciate the level of understanding of their readers provides us, the program directors in radiation therapy, radiation diagnosis, and nuclear medicine, with a major resource in furthering our residents' early training. In addition, the refreshing quality and clarity of this second edition will assure it a dominant position in our teaching libraries for years to come.

William T. Moss, M.D.
University of Oregon
Health Sciences Center
Portland, Oregon

Preface

This book is written to serve the same audience as its predecessor, people in the medical profession—radiologists, residents in radiology, medical students and technologists. It is a teaching tool, and as such, we expect it to be of special interest and particular value to those teaching radiation biology as well as those studying it. Furthermore, we believe it can have an appeal outside of medicine. Anyone wishing a basic understanding of some of the results of the interactions of ionizing radiations and ultrasound with living matter will find it useful.

Like its predecessor, it presumes the reader to have a foundation in basic physics. No substantial review is given. Also like its predecessor, it is not intended to be an exhaustive treatment of radiation biology or even of its medical aspects. Material and principles we deem of greatest relevance to medicine have been selected and presented in what we think is simple, straightforward prose and illustrations, accompanied by a number of representative references.

This book has much in common with the first edition, but there are several significant differences. It has been completely rewritten. Almost none of the original text and few of its illustrations have survived. This is not because of dissatisfaction with our earlier effort but because of two main factors: (1) an entirely new structure has been adopted and (2) much new information has been incorporated and outdated material has been eliminated.

This edition takes as its theme, dose. We begin by defining dose as a concept, follow this by discussing the absorption of radiant energy in matter and then by describing the biologic effects of increasing dose. The description of low-dose effects has been greatly expanded. Background information, permitting understanding of the production of these effects and their relevance to medicine, has been included. Thus, mutation, carcinogenesis, embryonic effects, and fetal effects are greatly stressed. Generally, and through medicine specifically, the human population is exposed to low doses of man-made ionizing radiations, and the potential consequences are currently of great concern in and out of the medical profession.

The new organization and lengthy development of the material on low-dose effects does not prevent our book from being eminently useful to residents in and practitioners of radiation therapy. Quite apart from the fact that low-dose effects are important considerations in radiation therapy, the book includes descriptions of effects produced by the therapeutic dose range and discussion of the problems presented.

We have tried to offer estimates of risks taken by persons receiving medical or occupational radiation. In itself this is a hazardous undertaking and, mindful of the controversies surrounding risk estimates, we have tried to describe and explain the bases of disagreement.

As for benefit of medical irradiation, we have spoken about it only in a general way. No specific benefit/risk evaluations have been made. Only persons trained in medicine, an area outside our expertise, can do this in individual cases.

Finally, no preface could be complete without an expression of our sincerest thanks to Constance Carvell who supervised the preparation of our manuscript. It was a big job, done expertly and with patient good humor, and we are grateful.

Donald J. Pizzarello
New York, New York

Richard L. Witcofski
Winston-Salem, North Carolina

Contents

1. The Elusive Concept of Dose 1

2. Matter and Energy Exchange by Ionizing Radiations 8

3. Mutation—An Effect on the Gene Pool 15

4. Mutation—The Effects of Irradiation on the Chromosomes 31

5. Radiation Carcinogenesis 42

6. Effects of Radiation on Embryonic and Fetal Development 65

7. Risks of Low-Dose Irradiation 75

8. Risks of Occupational Exposure 89

9. Cell-Survival Curves 94

10. Modifiers of Cellular Response to Irradiation 103

11. Repair and Recovery from Radiation Damage 119

12. Responses of Tissues and Organs to Irradiation 129

13. Responses to Total-Body Radiation 134

14. Risks of Exposure to High Doses of Radiation 146

15. Risks of Diagnostic Ultrasound 154

Index 159

Chapter 1

The Elusive Concept of Dose

1.1 INTRODUCTION

The intention of this book is to present what is known about the effects of exposing human beings to various radiations during medical procedures and as a result of their occupations. The book's concern is overwhelmingly with ionizing radiations, since these are the radiations most often used in medicine and to which many people are exposed occupationally. In recent years, increasing numbers of people have been exposed medically to ultrasound, and because of that, a short section has been added introducing the subject. As our book develops, it will become apparent that, for many effects, there is a high degree of uncertainty about the relationship of biologic response to the amount of radiation energy absorbed (dose). Following exposure to high levels of radiation, biologic responses often take a *fairly* predictable course. But most human beings are exposed medically and occupationally not to high but to low levels of radiation, often to very low levels. At present, the population of the United States receives an average yearly dose from medical radiations approximately equal to the average yearly dose from other sources of radiation, principally background exposure (Table 1.1). That received yearly

Table 1.1. *U.S. general population exposure estimates, BEIR 1980.*[6]

Source	Average Individual Dose (mrem/year)
Natural background (Cosmic, terrestrial, internal)	82
Medical	93[a]
X rays–79 *mrem/yr*	
Radiopharmaceuticals–14 *mrem/yr*	
Fallout (Weapons testing)	4–5
Consumer products	3–4
Nuclear industry	<1
Airline	0.61
Travel–0.6 *mrem/yr*	
Radiopharmaceutical transportation–0.01 *mrem/yr*	
TOTAL	185

[a]Note that the dose from medical exposure is 50% of the total.

from *all* other man-made sources combined is less than 10% of the natural background radiation.

Simply stated, since Roentgen's discovery of x rays in 1895, in about four human generations, the average yearly radiation exposure of the U.S. population has more than doubled, and nearly all the added radiation is from medical and dental *diagnostic* procedures.

Clearly, there is a great need to know what hazards, if any, are presented to people by this exposure and by exposures given during radiation therapy. Formidable difficulties stand in the way of collecting data that make possible unequivocal predictions of risks, especially in situations in which radiation exposure is at low levels. In this book we hope to present a summary of the existing factual information regarding biologic effects of medical radiation. We also hope to convey a sense of what is unknown and what remains to be learned, as well as a sense of prudent caution to medical users of radiation. We hope to make clear that often radiation risks are not precisely but only vaguely known, and/or may pertain to unborn generations.

1.2   THE ELUSIVE CONCEPT: DOSE

The idea that effects produced in living things by ionizing radiations are rooted in the amount of energy these radiations transfer to them appears to be so true that it seems unnecessary to state it.

The amount of energy transferred from radiations to living matter has come to be called "radiation dose," even when there is no therapeutic connotation. When people receive radiation as therapy for malignant disease, we speak of the dose given to treat the disease—just as we speak of the dose of drug needed to treat any other disease. But when people receive radiation for *diagnostic* purposes or are irradiated in the course of their work or because they live near a source of radiation, we continue to speak of their radiation "dose," despite the fact that no treatment is involved. The word has come to mean, simply, radiation energy absorbed.

Radiation dose is a concept which, at first, seems susceptible to precise definition. In fact, that is far from true, and has led to unreasonable expectations. Lay persons and even physicians who use radiations professionally ask and expect rather

precise answers to questions like: "As a physician, I often order diagnostic tests using radiation. *What dose* will my patient receive and what might it do?" This question has several answers, none of which is likely to be the expected or even a satisfying answer to the physician's query. There is no single dose from diagnostic procedures; instead, there are virtually an infinite number of doses received throughout the body, because during exposure, every part of the body receives a different dose. Is the physician interested in total body dose, marrow dose, gonadal dose, fetal dose? Each is different. Moreover, *response* to dose depends on a number of factors. Type of radiation used, rate of delivery, volume of tissue irradiated, proliferative status of tissue, age and sex of person irradiated—all, among other factors, influence radiation response, and not always to a precisely predictable degree. Yet, dose and effect are related, even if in a complex way. Radiations' effects do begin with energy absorbed in tissue (dose), and for that reason, we must begin by considering the units of dose and how they are measured.

1.3   SPECIFICATION OF RADIATION DOSE

Because nearly all radiation effects depend upon amount of energy absorbed, it is imperative to be able to measure dose in units that are precise and unequivocal. Units of radiation quantity must be specifically defined and universally accepted so that physical irradiation techniques and results can be reproduced.

Roentgen: the Unit of Exposure

The roentgen (R) was adopted in 1928 as a unit of exposure to medium-energy x radiation. The roentgen is the quantity of x or gamma rays that produces 2.58×10^{-4} coulombs/kg of air at standard conditions of temperature and pressure. In measuring the roentgen, an air volume of known size is irradiated, and the ions produced (electrical charge) are collected and measured. The choice of air as a standard substance was for convenience. Since air and water have an *effective* atomic number nearly the same as that of tissue, absorption of x-ray energy *per gram* of soft tissue, water, and air is almost the same even when x rays of widely varying wavelengths are used.

However, the roentgen had limitations as a unit of measure. By definition it was limited to x and gamma rays and did not include other types of radiation. Further, the definition of the roentgen holds only for lower energy radiations (up to 3 MeV).

Rad: the Unit of Absorbed Dose

It became desirable to introduce a unit that could be correlated more clearly to biologic and chemical effects than could the roentgen. The rad was a unit of absorbed dose that was not restricted to air and could be applied to all kinds of radiations. It is specifically defined as the deposition of 100 ergs/g in the medium of interest. As a general rule, the absorbed dose in soft tissue from 1 R of intermediate energy x or gamma rays is about 1 rad. The International Commission on Radiological Units[1] has recently proposed the adoption of Système Internationale (SI) units, in which the rad would be replaced by the Gray (Gy), which results in an absorbed energy 100 times greater than a rad (1 Gy = 100 rad = 1 joule/kg). Since the transition to SI units is expected to take some time, this book retains the units currently employed (rad, rem, curie).

Rem: the Dose Equivalent Unit

The rem was developed in response to a considerable body of evidence indicating that biologic effects per rad of various radiations are often different. The dose equivalent (DE) is defined as the absorbed dose adjusted by a term that expresses the differences in biologic effectiveness of various types of radiation as compared to ordinary x rays. This term is called the "quality factor" (QF) when used in personnel protection and as "relative biological effectiveness" (RBE) in radiobiology. The appropriate SI unit for dose equivalent is the Sievert (Sv), which is related as 1 Sv = 100 rem.

1.4 DOSE IN DIAGNOSTIC ROENTGENOLOGY

Since in diagnostic studies only a small portion of the body is exposed to a highly restricted (collimated) beam of radiation, regions of the body within the beam receive the highest dose. In general, radiation dose (the absorbed energy) changes as radiations move through tissue, the greatest being at or near the point of entry of the radiations and the least being at the point most distant, for tissue directly in the beam (Fig. 1.1). Because diagnostic x rays are rapidly absorbed by body tissues, organ doses at a depth are usually considerably less than surface or skin doses. For example, because of the posterior position of the spine, a posterior-anterior view of the lumbosacral spine results in a mean marrow dose three times greater than that yielded by the comparable anterior-posterior view. Also, dose varies depending upon the capacity of various structures within an irradiated volume to absorb radiation. Bone differs significantly from soft tissues in its capacity for absorbing lower-energy x rays and consequently receives or absorbs a higher dose than soft tissues from exposure to a given quantity of these radiations. How then do we respond to the most common question about an x-ray examination: "How much radiation is received from this examination?" At present, there is *no* uniform convention to answer this question. A variety of ways have been used, including: exposure, entrance skin dose, bone marrow dose, organ of interest dose, midline dose, and integral dose.

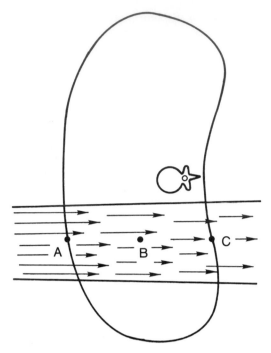

Fig. 1.1. A diagrammatic representation of the absorption of a diagnostic x-ray beam in a patient's body. The point (A) on the surface facing the x-ray tube would receive the highest dose. Assuming a patient thickness of 20 cm, the dose at the midplane (B) would be approximately 10 to 20% of that at A, while the dose at C would be about 1 to 10% of that at A.

Exposure in Air at the Entrance Skin Surface Measured in R

This measurement is obtained by placing an ionization chamber at the position where radiation enters the patient's body. It is widely used because it is easily obtained, but it also gives the least information.

Entrance Skin Dose

This includes radiation scattered to the skin surface and is usually calculated from exposure at that point. Because of the rapid drop in dose with depth in the patient's body, its value is ordinarily much higher than that for more "sensitive" organs at a depth.

Bone Marrow Dose

The dose to active bone marrow at a particular point may be determined, but more commonly, "mean marrow" dose is calculated. To do this, absorbed doses to tissue at various bone marrow sites are determined and weighted for the fraction of active marrow they represent. Then the weighted doses to all irradiated sites are summed to give the mean active marrow dose averaged over the whole marrow. Since active marrow is distributed approximately as: pelvis 36%; spine 28%; skull 13%; and ribs and sternum 10%, procedures such as the barium enema, IV pyelogram, and lumbar spine radiography result in the highest mean marrow doses.

Organ-of-Interest Dose

This includes dose levels to such organs as gonads, lens of the eye, and thyroid gland, which are often too complicated to measure on individual patients. Methods of calculation are available for these and other internal organs.[2] Organs near the tissue volume directly irradiated may also receive a significant dose through scattered radiation.

Midline or Midplane Dose

To obtain an accurate value one must have detailed knowledge of intervening tissues and beam quality as well as entrance dose.

Integral Dose

This is a measure of total energy absorbed within an irradiated volume. The mass of tissue irradiated multiplied by the average dose gives a gram-rad. Because this is a measure of total energy deposited in the patient, some authorities feel it may be of great importance in estimating risks of cancer induction.

Radiation Doses from Radiography

Estimates of radiation dose levels from common radiographic studies are given in Table 1.2 for skin, bone marrow, and gonads. The values in the table should not be considered to apply accurately in individual cases. They are mean values, and

Table 1.2. *Estimated doses in rads for radiographic examinations*[2-6]

Type of Examination	Mean Skin Dose per Film	Mean Marrow Dose per Exam*	Mean Gonadal Dose per Exam† Male	Female
Skull	0.3	0.08	—	—
Skull CT scan	5.0	0.3	—	—
Cervical spine	0.25	0.05	—	—
Chest	0.05	0.01	—	—
Mammography	1.3	—	—	—
Thoracic spine	1.3	0.24	—	—
Upper GI series	0.8	0.54	—	0.17 (0.3)
Barium enema	0.63	0.88	0.18 (0.3)	0.90 (1.7)
Cholecystography or cholangiogram	0.77	0.17	—	0.08
IV pyelogram	0.5	0.42	0.21	0.59
Abdomen, KUB	0.77	0.15	0.1	0.22
Lumbar spine	2.5	0.45	0.22	0.72
Pelvis	0.5	0.09	0.36	0.21
Hip	1.0	0.07	0.60	0.12
Upper extremity	0.13	—	—	—
Lower extremity	0.11	0.02	—	—

—Less than 0.01 rad

*"Mean marrow dose" is the dose received by any portion of the active marrow averaged over the whole mass of active marrow to compensate for partial organ irradiation.

†The doses listed are for radiography. The numbers in parentheses are estimates of the total gonadal dose from radiography and fluoroscopy.

there can be considerable variability. Note that gonadal doses are highest from procedures that directly irradiate the pelvis. The genetic impact on the population, however, depends not only upon the gonadal dose per examination, but also on the frequency of the examination, as well as the distribution of the ages of patients on which the examination is performed. For example, a given procedure may yield a relatively high gonadal dose, but if it is rarely used, may have relatively small impact on the genes of a population. This concept is dealt with in greater detail in Chapter 3.

1.5   DOSE IN NUCLEAR MEDICINE

In nuclear medicine procedures, patients are irradiated by radioactive materials localized in certain organs or distributed throughout their bodies. Because radionuclides are taken internally, there are many variables, and dose to individual patients cannot be measured but only estimated by calculation. Numerous factors produce uncertainties in these calculations.[7-10] Some follow.

Activity, Route of Administration, State, and Type of Radiation

Factors that affect radiation dose include activity of administered radiopharmaceuticals and their half-lives; chemical and physical state of the radiopharmaceutical; route of administration; and type of radiation emitted and its energy. Half-life, emitted radiation, and energy usually are known with great precision.

Redistribution of Radionuclides

In nuclear medicine, radiopharmaceuticals are usually introduced into blood either by intravenous injection or by absorption from the gastrointestinal tract. They are distributed rapidly throughout the plasma volume and from plasma to other body "compartments." They may be concentrated for a time in one or more organs, but eventually they are eliminated by a combination of two processes: biologic excretion and radioactive decay.

The *rate* of radioactive decay is almost always well known and is given by radionuclides' physical half-lives. But *rate* of biologic elimination, which also has a half-life known as biologic half-life, is rarely well known. Both processes decrease the amount of radioactivity in the body and combine to give what is called an "effective half-life," a time less than either the physical or biologic half-lives. However, there are scant data on biologic elimination of most commonly used radiopharmaceuticals, and there is great variability in elimination rates even among well individuals. But many individuals studied in nuclear medicine are ill, and pathologic states can have a profound effect on biologic handling and excretion of radiopharmaceuticals.

Pathology

Most available information on distribution of internally administered radiopharmaceuticals does not take into consideration the effects of various illnesses. In nuclear medicine it is the rule rather than the exception to have pathologic changes in the organ under study as well as other organs important in radionuclide distribution. Under these circumstances, significant deviations from normal distribution occur. For example, colloidal ^{99m}Tc is usually distributed with approximately 90% in liver and the remainder divided between spleen and bone marrow. However, a diseased liver may concentrate only 10% of a given dose. The remainder continues to be shared between spleen and bone marrow, but the concentration in the spleen, because it is smaller than the liver, may actually exceed that in liver. Another example is treatment of toxic goiter. Higher doses of radioiodine (^{131}I) are often necessary to adequately treat toxic nodular goiter than diffuse toxic goiter. Most damage produced by ^{131}I results from its beta particles,

which have a maximum penetration in soft tissue of about 2 mm. In the diffuse toxic gland, uptake of ^{131}I and radiation dose is relatively uniform, but in toxic nodular goiter, distribution and uptake are irregular. The most active areas receive high doses, but many inactive nodules receive little. Owing to "sparing" of inactive areas, adequate thyroid function is preserved and a low incidence of myxedema results.

Age of the Patient

Variation in dose depends on the patient's size and age. Not only are organs smaller in infants and children than in adults, but body proportions are also different. Infants and children receive greater radiation doses than adults from given quantities of radionuclides, and their tissues may be more susceptible to deleterious effects of ionizing radiations. Little is known of biologic half-life in children because of proper reluctance of investigators to give them radioactive materials to study normal physiologic processes. In clinical nuclear medicine, smaller quantities of radioactivity are used in young patients than in adults. Yet reduction in amount based only on relative weights may result in a poor estimate of radiation risk.

Radiation Doses from Radiopharmaceuticals

Typical examples of dose levels from radiopharmaceuticals widely used for diagnostic studies are given in Table 1.3. The table lists dose to the "critical organ" (the organ receiving the highest dose) and the gonads. In nuclear medicine procedures, whole-body doses are more uniform and usually higher than in plain radiographic examinations but generally less than 0.2 rad.

SUMMARY

1. At present, the population of the United States receives an average

Table 1.3. *Typical radiation doses from selected commonly used radiopharmaceuticals*

Radiopharmaceutical	Administered Dose (mCi)	Dose to Critical Organ (Rad)	Dose to Gonads* (Rad)	
^{99m}Tc-Pertechnetate	10	1.3 (thyroid)	0.09	(M)
			0.3	(F)
^{99m}Tc-DTPA	10	4.5 (bladder)	0.15	
^{99m}Tc-Sulfur colloid	3	1.0 (liver)	0.3	
^{99m}Tc-Polyphosphate	10	3.1 (bone)	0.15	
^{99m}Tc-Albumin	2	0.1 (blood)	0.08	
^{99m}Tc-MAA	3	0.3 (lung)	0.03	
^{131}I-MAA	0.3	2.0 (lung)	0.1	
^{131}I-Sodium iodide	0.05	50 (thyroid)	0.004	(M)
			0.007	(F)
^{125}I-Sodium iodide	0.05	25 (thyroid)	0.015	(F)
			0.009	(M)
^{131}I-Rose bengal	0.15	5.3 (large intestine)	0.2	
^{131}I-Hippuran	0.2	0.2 (kidney)	0.006	
^{67}Ga-Citrate	3	2.7 (large intestine)	0.8	
^{75}Se-Selenomethionine	0.2	5.0 (liver)	2.2	(M)
			1.0	(F)
^{51}Cr-Sodium chromate	0.05	1.6 (spleen)	0.01	
^{201}Tl-Chloride	1.5	1.8 (kidney)	0.9	
^{133}Xe-Xenon	10	0.2 (lung)	0.001	

All of the above procedures result in whole-body doses of less than 0.2 rad except ^{67}Ga-citrate, ^{201}Tl-chloride, and ^{75}Se-selenomethionine, which deliver doses of 0.8, 0.4, and 1.6 rad respectively.

*Doses for both testes and ovaries are given when they are quite different.

yearly dose from medical radiations (93 mrem/year) that is approximately equal to the average yearly exposure to other sources of radiation.

2. It is difficult to give precise values for the radiation doses associated with diagnostic radiologic and nuclear medicine procedures.

REFERENCES

1. International Commission on Radiation Units and Measurements (ICRU) Report 33. Radiation Quantities and Units. Washington, D.C., April 1980.
2. Rosenstein, M.: Organ Doses in Diagnostic Radiology. HEW Publication FDA 76-8030, U.S. Government Printing Office, Washington, D.C., May 1976.
3. Gonad Doses and Genetically Significant Dose from Diagnostic Radiology U.S., 1964 and 1970. HEW Publication FDA 76-8034, U.S. Government Printing Office, Washington, D.C., April 1976.
4. Antoku, S., and Russel, W.J.: Dose to the active bone marrow, gonads, and skin from roentgen-ography and fluoroscopy. Radiology, *101*:669–678, 1971.
5. Shleien, B., Tuck, T.T., and Johnson, D.W.: The mean active bone marrow dose to the adult population of the United States from diagnostic radiology. Health Phys., *34*:587–601, 1978.
6. National Research Council, Advisory Committee on the Biological Effects of Ionizing Radiations (BEIR III): The Effects on Populations of Exposure to Low Levels of Ionizing Radiation. National Academy of Sciences, Washington, D.C., 1980. (See particularly Chapter III, Sources and rates of radiation exposure in the United States).
7. International Commission on Radiation Units and Measurements (ICRU) Report 32: Methods of Assessment of Absorbed Dose in Clinical Use of Radionuclides. Washington, D.C., November 1979.
8. Dillman, L.T., and von der Lage, F.C.: Radionuclide Decay Schemes and Nuclear Parameters for Use in Radiation-Dose Estimation. MIRD Pamphlet No. 10. New York, The Society of Nuclear Medicine, 1975.
9. Snyder, W.S., Ford, M.R., and Watson, S.B.: "S," Absorbed Dose per Unit Cumulated Activity for Selected Radionuclides and Organs. MIRD Pamphlet No. 11. New York, The Society of Nuclear Medicine, 1975.
10. Radiopharmaceutical Dosimetry Symposium HEW Publication FDA 76-8044, U.S. Government Printing Office, Washington, D.C., 1976.

Chapter 2

Matter and Energy Exchange by Ionizing Radiations

2.1 INTRODUCTION

Detectable injury or damage to living things as a result of exposure to ionizing radiation is the result of a long, complex chain of events. The first of these is the transfer of energy from ionizing radiation into the matter of living things. The *mode* of energy transfer characteristic of ionizing radiation is the production in matter of excited and ionized atoms or molecules.

2.2 ATOMIC STRUCTURE

Energy from radiation is absorbed in matter by interactions at the atomic level. Accordingly, some understanding of atomic structure is necessary for an understanding of radiation interactions and radiation damage to living things. Each atom consists of a small dense nucleus which contains positively charged protons and uncharged neutrons and includes practically the whole mass of the atom. The radius of atomic nuclei does not exceed 10^{-12} cm.

The atomic nucleus is surrounded by a cloud of small, moving electrons which travel in orbits extending to a radius of about 10^{-8} cm. The electron cloud has little mass compared to that of the nucleus, but it occupies a great deal of space and gives the atom its size. Because of the cloud's diffuse nature, the atom is mainly empty space. In fact, if an atom were enlarged to the size of an ordinary room, the nucleus would occupy a space about the size of the period at the end of this sentence. Because of the "emptiness" of matter, ionizing radiations may pass through many atoms before interacting with any of them.

Each electron carries a unit negative charge. The cloud of electrons holds together because the electrons are moving and because a powerful electric attraction is exerted by the positive charge of the nucleus. The attraction of the nucleus predominates over repulsion between electrons because the charge on the nucleus is greater.

Each electron rotates about the nucleus in a discrete orbit which is at a particular distance from the nucleus. Electrons in inner orbits are more tightly bound than those in outer orbits. Atoms which have high atomic numbers (the atomic number of an atom is equal to the number of protons in its nucleus and is called the "Z value") attract electrons in any given orbit more strongly than do electrons in the same orbit of an element with a low atomic number. In the stable configuration, the atom is electrically neutral; that is, the number of protons in the nucleus is equal to the number of electrons in orbit.

2.3 IONIZATION

The principal means by which ionizing radiations dissipate their energy in matter is by ejection of orbital electrons. The ejection of one or more is called ionization.

Atoms are electrically neutral, but when they are ionized, the loss of orbital electrons leaves them positively charged. An ionized atom and a dislodged electron constitute an ion pair.

2.4 EXCITATION

Not every interaction of an ionizing radiation in matter results in ionization. Excitation accounts for a significant percentage of energy dissipated by ionizing radiations in tissue. The process involves electrons, called "valence electrons," in the outermost orbit. Since they are at the greatest distance from the nucleus, they are relatively loosely bound. If a small amount of energy is given them, they become excited; that is, they overcome the energy that binds them in their orbit and move farther away from the nucleus. The orbits to which they move are usually empty and are called "optical orbits."

An excitation of outer orbital electrons brings about a change in the chemical forces that bind atoms within molecules. This change of forces may or may not lead to a regrouping of affected atoms and a different molecular arrangement. Thus, excitation of outer orbital electrons appears to be an indirect method of inducing chemical changes by supplying needed energy and

breaking bonds between the atoms of molecules.

2.5 FREE RADICALS

Ions are extremely transient, having lifetimes of the order of 10^{-10} seconds, and ions or *excited* atoms or molecules may give rise to another unstable chemical species, free radicals. Free radicals are atoms or molecules which are characterized by an electron orbit in which there is a *single* unpaired electron with respect to spin. In atomic orbits, electrons rotate on their axes. All orbits should have pairs of electrons axially rotating in opposite directions for stability. When that condition does not obtain and there is an unpaired electron, the entity is a free radical. Atomic hydrogen, with a single proton in its nucleus and a single electron in its inner orbit, is an example and is, in fact, the simplest free radical.

Free radicals have a relatively high likelihood of reacting with other molecules. They may achieve stability by "pairing" their unpaired electron with orbital electrons of the molecules with which they react. While they are unstable and do react with other molecules, they are considerably more stable than ions (lifetimes of the order of 10^{-5} seconds for free radicals as opposed to 10^{-10} seconds for ions). Because of this, they persist much longer than radiation-produced ions. Small free radicals can diffuse through the medium they are in before reacting with a molecule that provides them with the needed electron. This is one way by which small free radicals affect molecules distant from where the free radicals were formed. It must be clear, however, that since the lifetime of free radicals is about 10^{-5} seconds, they will not diffuse very far from the place of formation, as there is too little time. Individual free radicals undergo some kind of change or reaction either at or within a small fraction of a micron from the point of formation.

There is another way, however, for radiation-produced free radicals to affect other molecules at a distance. They may transfer their "extra" electron to a nearby molecule, which in turn, may pass it on. The electron can be passed through a succession of molecules. Each receiving it becomes a free radical. As soon as the electron is passed on, the molecule is no longer a free radical. This chain reaction may occur in a large number of molecules and, in a manner not understood, results in critical changes in organic molecules.

Similarly, a free radical may "snatch" an electron from a neighbor, which then becomes a free radical. A chain of electron "snatchings" may ensue and again, in an unknown manner, an organic molecule may be critically changed in the process.

Radiation damage to organic molecules occurs almost exclusively by free-radical interactions and not by the ions produced by radiations. The lifetimes of these ions are so short that they rarely react directly to change organic molecules. Before that can happen, they produce free radicals, and these, with their longer lifetimes and ability to chain react, damage biologically important organic molecules. Ionizing radiation produces free radicals in abundance by ejecting electrons from molecular orbits or by causing electron rearrangement by excitation. Both actions leave orbits with electron deficits and unpaired electrons.

In cells, water is the most abundant source of radiation-produced free radicals. Excitation causes molecules, such as water, to dissociate yielding hydrogen and hydroxyl-free radicals.

$$H_2O \xrightarrow{\text{energy}} H\cdot + OH\cdot$$

Ionization of water can also lead to formation of the same free radicals. These may interact with other cellular molecules and, by sharing an electron, become part of one of the molecules. When this occurs, a critical, discrete change, at that site on the molecule or in the molecule generally, may occur. If the molecule is a key molecule in cellular metabolism, interactions at critical sites or general changes in the molecule may interfere with the normal function of affected cells.

2.6 TYPES OF IONIZING RADIATIONS

Ionizing radiations may be classified according to their origin (the products of radioactive decay, x-ray machines, particle bombardment, or nuclear reactors) or, more commonly, according to their physical properties. They fall into two general categories—those which have mass (corpuscular or particulate) and those which are energy only (nonparticulate or electromagnetic). Those with mass may be charged or uncharged, but nonparticulate radiations are never charged. The characteristics and origins of some types of ionizing radiations are summarized in Table 2.1.

2.7 THE INTERACTION OF X RAYS AND GAMMA RAYS

X-ray and gamma-ray photons are electromagnetic radiations which have no mass or charge. In their interactions with matter, the energy of photons is transferred by collision, usually with an orbital electron in an atom of the absorbing medium. Following such a collision, an electron may either have been moved to an orbit more distant from the nucleus (the atom is excited) or, more commonly, it will have been ejected from the atom (ionization) with high energy and at a high speed. The energy given to the electron will be dissipated as it moves through the medium; it will ionize and excite atoms with which it interacts. Thus, the damaging actions of x rays and gamma rays are exerted almost entirely through secondary fast electrons ejected from atoms.

Since the interaction of x-ray and gamma-

Table 2.1. *Types of ionizing radiation*

Type	Mass	Charge	Description	Produced by
Alpha	4	+2	Doubly ionized helium atom	Radioactive decay primarily of heavy atoms
Beta (negatron)	0.00055	−1	Negative electron	Radioactive decay and betatrons
Beta (positron)	0.00055	+1	Positive electron	Radioactive decay and pair production
Protons	1	+1	Hydrogen nuclei	Van de Graaff generators and cyclotrons
Negative π mesons	0.15	−1	Negatively charged particle with a mass 273 times an electron	Accelerators
Heavy nuclei	Have a range of masses	Have a range of charges	Any atom stripped of one or more electrons and accelerated will be an ionizing particle. Deuterons and carbon atoms are examples.	Accelerators
Neutrons	1	0	Neutral	Atomic reactors, cyclotrons
Gamma rays	0	0	Electromagnetic radiation	Radioactive decay
X rays	0	0	Electromagnetic radiation	X-ray machines and from the rearrangement of orbital electrons

ray photons with matter depends upon chance collision with electrons, these radiations are capable of penetrating deeply into matter and passing vast distances through it without having interacted.

2.8 INTERACTION OF PARTICULATE RADIATIONS WITH MATTER

Radiobiologically, the interactions of charged particles with living matter are important because they are directly responsible for the damage we observe. These interactions are distinct from those of x and gamma rays, because particulate radiations may have both mass and charge. These properties make possible interactions with matter not only by direct collision with electrons in orbits of atoms but also by interactions (this can be attraction or repulsion) between the charge of particulate radiations and that of orbital electrons. However, the final result of both electromagnetic and particulate radiations will be

the same; that is, ionization, the production in matter of high-speed electrons.

Neutron (uncharged particles) interactions depend upon chance collisions with atoms and they penetrate great distances in matter of all kinds. They are classified by their only distinctive property—their energy. Depending on that factor, they are usually considered as *fast* or *slow*. Fast neutrons lose energy mainly by collision with atomic nuclei, and are of the type which are of interest to radiotherapy. For practical purposes the major mode of energy loss of fast neutrons in soft tissue may be considered the *ejection* of high-speed protons (hydrogen nuclei). Soft tissue has many hydrogen nuclei because it is principally water. The protons have a variety of energies depending upon the neutron energies, but all of them are highly ionizing particles with a high LET.

2.9 THE EFFECT OF PARTICLE CHARGE

The effect which one charged body has upon another is related to the distance between them and to the amount of charge on each. As the distance between charged bodies increases and/or their charges decrease, the effect they have on each other diminishes. The rate at which energy is lost along the track of a charged particle depends upon the square of the charge on the particle. The more highly charged the particle, the more intense its electric force field and the greater the likelihood of its producing ionization and excitation of atoms along its track. The influence of particle charge is to increase the number of interparticulate interactions and hence to produce more ion pairs *per unit of path length*.

2.10 THE EFFECT OF PARTICLE VELOCITY

Two particles with the same energy do not necessarily have the same velocity. For example, protons, because their rest mass is about 2000 times that of electrons, are moving much slower than electrons when the kinetic energies of both are the same. Particle velocity influences interactions between charged particles, because velocity controls the length of time the electric force field of the particle is exerted in any given place. The effect (attraction or repulsion of orbital electrons) by the force field of the charged particles depends on how long the force is applied. Slower moving charged particles produce more ionizations per unit of path length than faster moving ones. Two particles having the same kinetic energy may produce vastly different ionization density along their tracks if their masses differ. For example, even though their kinetic energies are the same, a 1-MeV proton would produce an ionization density 100 times greater than a 1-MeV electron; the more massive proton would be moving more slowly than the electron (ionization density depends inversely on the velocity of the charged particle).

2.11 LINEAR ENERGY TRANSFER (LET)

Specific ionization, the number of ions formed per unit length of track traversed, takes into consideration energy transferred to the medium only by means of ionization; but energy will also be transferred to the medium by excitation. At present the full biologic significance of excitation is not known, but a substantial amount of energy is transferred to tissue in this way. A unit which has been devised to account for all the energy liberated along the path of an ionizing particle, irrespective of the mechanism, is Linear Energy Transfer (LET). LET is the energy released (usually in keV) per micron of medium (tissue) along the track of any ionizing particle.

Since rate of loss of energy by ionizing particles will be affected by the velocity (V) and the charge (Q) on such a particle (Q^2/V^2), a relatively slow moving highly charged particle will have a high LET. A faster moving particle and/or one with a lesser charge will have a much smaller LET. Biologic damage is related to LET. In general, particles with high LET are more likely to produce change in a given volume of living matter, because their interactions are produced more closely together than are those of low-LET radiations.

The increase in biologic damage with increasing LET does not, however, continue indefinitely. With radiations of extremely high LET, more energy is transferred to the system than is needed to produce even the maximum biologic effect—death. Such radiations will "overkill," that is, they will deposit more energy in cells or tissues than is required to kill them.

LET is not a static or constant value but is different even for the same particle over different portions of the track. This is so because, while charge on a particle is a constant factor, the velocity is continually changing (decreasing) along the particle track. Each interaction (excitation or ionization) involves a loss of energy from the particle and a concomitant deceleration. As a result, LET gradually increases along a particle track with a very dramatic increase occurring just before the particle comes to rest. This peak in rate of energy dissipation is called the "Bragg peak" (Fig. 2.1). Typical LET values are given in Table 2.2.

Particles with different degrees of charge produce different tracks. Highly charged particles interact frequently; consequently, they have a high LET, and the ionizations along their tracks are very dense. Particles with lesser degrees of charge are sparsely ionizing; they have a lower LET.

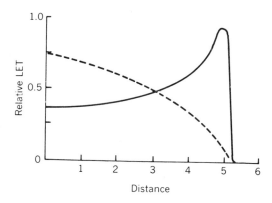

Fig. 2.1. The rate of dissipation of energy of a charged particle as it moves along its path. The rate increases gradually as the particle slows down but dramatically increases (the "Bragg peak") just as the particle is coming to rest. The dotted line indicates the residual energy of the particle as it moves along its track.

2.12 RBE OR QUALITY FACTOR

Ordinarily we think that biologic response to radiations depends mainly on dose. The biologic effectiveness, the efficiency of a particular radiation for producing a given effect, however, varies. Measures or estimates of this variation are called RBE (relative biologic effectiveness) values. The variation is due to the fact that effects of ionizing radiation depend not only on the total dose administered over a given period of time, but also on the distribution of that energy. Usually, the distribution of energy along individual tracks (LET) will determine the effectiveness of a radiation for producing a given result. For most radiations, RBE increases as the LET increases; a given dose of high-LET radiations will be more efficient than the same dose of sparsely ionizing radiations.

The degree of difference can be considerable (Fig. 2.2). In most RBE determinations, the standard radiations against which others are compared are medium-energy x rays (200 kvp with a half-value layer of 1.5 mm of copper).

The RBE has its greatest meaning when it is applied to small volumes of test material, such as bacteria or tissue culture cells.

Table 2.2. *LET values of ionizing particles*

Particle	Charge	Energy (MeV)	LET (keV/μ)
Electron	−1	0.001	12.3
		0.01	2.3
		0.1	0.42
		1	0.25
		200 kvp x rays*	0.4–36
		Cobalt 60γ rays*	0.2–2
Proton	+1	Small	92
		2	16
		5	8
		10	4
Alpha	+2	Small	260
		3.4	140
		5	95
Neutron**		2.5	15–80 (peak at 20)
		14.1	3–30 (peak at 7)

*This applies to the LET of secondary electrons ejected by photons.

**Neutrons produce no ionization directly in passing through tissue. These values are for the protons ejected in collisions.

SUMMARY

1. The atom consists of a massive portion, the nucleus, which contains the

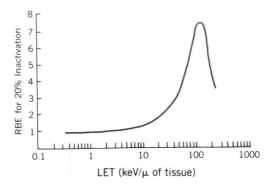

Fig. 2.2. An example of variation of RBE with LET for a particular experiment. The end point for biologic effect is the ability of the radiation to inactivate 20% of the cells growing in a tissue culture. The fall in RBE at the highest LET values is a result of the transfer of more energy than necessary to inactivate sensitive sites in the cells; hence, there is a drop in efficiency. (From Barendsen, G.W., and Walter, H.M.D.[2])

positive protons and uncharged neutrons.

2. The nucleus is surrounded by negatively charged electrons which are found in discrete orbits around the nucleus.

3. In the stable atom, protons and electrons are equal in number—the atom is electrically neutral.

4. Radiations may have sufficient energy to remove an electron from an atom and produce an electric charge (ionization) or, with less energy, to move an electron to an orbit farther from the nucleus (excitation).

5. These processes, particularly ionization, are responsible for biologic damage produced by ionizing radiations.

6. A major part of the damage to the important key molecules of cells probably results from chemically active free radicals resulting from excitation and ionization.

7. X rays and gamma rays lose their energy principally by ejecting high-speed electrons from atoms. These secondary electrons dissipate their energy by the production of further ionization and excitations.

8. LET is a measure of energy lost per unit of path length (keV/μ) as particles lose energy in matter (tissue).

9. LET varies as the square of the charge on a particle and inversely as the square of its velocity.

10. Charged particles, such as protons, alpha particles, and heavy nuclei, lose most of their energy through electron interaction but may also interact with nuclei.

11. High-speed neutrons lose their energy in tissue primarily by the ejection of high-LET protons from water.

12. The roentgen is the unit of exposure for x rays and gamma rays.

13. The rad is the unit of dose which relates to energy absorbed.

14. RBE (relative biologic effectiveness) relates energy absorbed to effectiveness in production of end point (rads to biologic effect).

REFERENCES

1. Rossi, H.H.: The role of microdosimetry in radiobiology. Radiat. Environ. Biophys., 17:29–40, 1979.
2. Barendsen, G.W., and Walter, H.M.D.: Effects of different ionizing radiations on human cells in tissue culture. Radiat. Res., *18*:106, 1963.
3. International Commission on Radiation Units and Measurements (ICRU) Report 10a, NBS Handbook 84. Radiation quantities and units. Washington, D.C., 1962.
4. Pizzarello, D.J., and Witcofski, R.L.: Basic Radiation Biology. Philadelphia, Lea & Febiger, 1975, Chapters 1–3.

Chapter 3

Mutation—An Effect on the Gene Pool

3.1 INTRODUCTION

One of the possible consequences of exposure of the human population to low doses of ionizing radiation that has occupied the thought and concern of many people is an adverse effect on the population of the future.

This could happen (1) if radiation exposure of people living now causes significant numbers of deleterious mutations in their gametes and (2) if those mutations were put in the gene pool, i.e., transmitted to descendant generations. Should those events happen, the possibility arises of people not yet born inheriting some defect of metabolism or function which could decrease their ability to compete and survive. Their viability, their capacity for survival, might be impaired and their life span accordingly shortened. In more concrete terms, a possible consequence of the use of ionizing radiation to benefit people living today might be an injury to people yet to be born—and a shortening of the average life expectancy of future populations.

Because the use of radiations in medicine is widespread, at least in the so-called "developed nations of the world," it is considered a problem of considerable potential importance.

3.2 MUTATION

Mutation is a natural phenomenon that occurs in all living populations. DNA, the material of which genes are made, is a stable molecule, generally transmitted from one generation to the next largely unchanged. However, it is not absolutely stable; for unknown reasons small changes occasionally occur in its structure and composition. These changes are called "spontaneous mutations." Because they usually are small and do not involve large segments of the genome, they are known as "point mutations."

3.3 SPONTANEOUS MUTATION FREQUENCY

The probability that any given gene will mutate spontaneously is small. In humans the spontaneous frequency is estimated to be approximately 10^{-5} (or less) per gene per generation.[1,2] This means that, in a human generation (about 30 years), the chance that any given gene will mutate is about 1 in 100,000. There are, of course, many thousands of genes in every cell and many cells in every body, so the chance of one or few mutations occurring within a generation is good. The mutation of a given gene in a given generation has a low probability of occurrence; that some gene will mutate in a given generation is likely.

3.4 RADIATION-INDUCED MUTATIONS

Exposure to ionizing radiations increases the mutation frequency. It appears that ionizing radiations are mutagenic; they can induce mutations. It is important to stress that radiogenic mutations are not unique and do not differ from those that occur spontaneously. The effect of radiation is not to produce hitherto unknown or different mutations, but to increase the frequency with which the spontaneous ones occur.

3.5 THE ELIMINATION OF MUTATIONS

Point mutations (spontaneous and/or radiogenic) may be beneficial (produce a better "fit" with the environment than the nonmutant form), neutral (do not affect the "fit" with the environment) or deleterious (produce a lesser "fit" with the environment than the nonmutant form). Population and environmental pressure develop against deleterious mutations and they are slowly eliminated from the gene pool. The mechanism is simple. The lesser fit they have with the environment impairs the viability of those expressing them and shortens their lives. If they are sufficiently deleterious, they can shorten life so much that individuals inheriting them may not live long enough to reproduce and the mutation dies out with them.

Because the process of elimination rids populations of genes that weaken them, it is good in that respect. However, the elimination of unfavorable mutations is usually accompanied by some degree of pain and suffering on the part of those expressing them. Their life is shortened and the metabolic and/or physical abnormalities deriv-

ing from the mutation can cause physical and psychic pain.

Mutations enter the gene pool spontaneously at low rates (for humans, in the region of 10^{-5}/gene/generation) and deleterious ones leave it also at low rates. The rate of entry and elimination are, in undisturbed populations, about equal, resulting in an equilibrium. In turn, the equilibrium results in a more-or-less constant but small fraction of the gene pool being mutant. That fraction is called the population's or gene pool's mutation burden or load.

3.6 CHANGES IN MUTATION FREQUENCY

Spontaneous mutation frequency is probably not an absolutely fixed quantity. In the natural course of events, things occur which may transiently raise or lower it. For example, exposure to ionizing radiation increases the frequency of mutations. But *background radiation,* a presumed source of about 1 to 6% of all spontaneous mutations,[1] changes. Changes in quantity of cosmic radiation occur when solar storms and other celestial events send more radiation to Earth. On Earth itself, background radiation varies according to the composition of the crust in various locales. There are variations according to altitude. All these cause fluctuations in mutation frequency so that such fluctuations *per se* are not necessarily cause for concern if increases in mutation frequency are offset by later declines (as, for example, may happen when sunspot activity increases and decreases, transiently changing the background radiation).

3.7 CHANGES IN MUTATION ELIMINATION RATE

It is to the advantage of populations to keep mutant gene forms in the gene pool at all times. Even mutations that prove deleterious when they occur might be beneficial should a change occur for which they happen to be suited. This is a paradox. Deleterious mutations, because of the pressures against them, compete less well and go out of existence. Yet their presence in the gene pool may be potentially desirable. The paradox is resolved in the following way. Most point mutations are at first recessive, and because of that, can remain in gene pools for many generations, masked by the nonmutant forms that dominate. Various cellular functions and characteristics are controlled by *two* bits of genetic matter (called alleles) on two different chromosomes. When the chromosomes of cells are examined carefully, it becomes clear that every chromosome in each cell is a member of a pair. Chromosomes come in sets, and there are two of them, one set contributed by the father and one by the mother when sperm and egg fuse at fertilization. The sets are called "homologs" of each other, and each chromosome in one set has a homologous partner in the other set.

In the normal course of events, when spontaneous point mutations occur, they happen to one allele of a gene but not to the other. The same holds for radiation-produced mutations. Radiations passing through cells may produce a mutation in one allele on one chromosome, but it is unlikely that they will produce exactly the same change in the other allele which is on another chromosome in that nucleus (Fig. 3.1).

When both alleles of a gene are alike, they are said to be homozygous; if they differ, they are called heterozygous. When alleles are present homozygously, only one genic expression is possible. When alleles are present heterozygously, two genic expressions are possible, but usually one only is expressed, and that one is called the dominant one. Frequently, it is the mutant gene forms that are not expressed, and these are called recessive. Consequently,

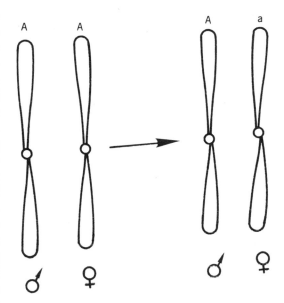

Fig. 3.1. Homologous chromosomes with the alleles A and A occupying the same locus on each (left). A mutation occurs such that A on one becomes a. The new genotype is Aa.

after mutation, the character produced by the dominant (nonmutant) partner is the one generally expressed. In some cases, however, the presence of the recessive allele does lessen the level of function of the dominant, and recessives are not always unfelt. The genotype changes as a result of mutation, but quite often the *functioning* of the mutant cell (the phenotype) remains the same. Since dominant homozygous genotypes and dominant heterozygous genotypes are usually expressed in about the same way, neither is a better or lesser fit for the environment than the other; neither has any advantage.

The deleterious nature of such mutations is most likely to be fully expressed only when they occur together, homozygously, in the cells of individuals. Then, the impairment of function they cause may shorten the life expectancy. If life expectancy is shortened enough that it drastically reduces the probability of reproduction, the mutant will not reenter the gene pool.

The probability that two mutants will occur homozygously is remote at any given time because the number of mutants is small relative to the number of nonmutants. In the normal course of events, they will be paired with and masked by a dominant in heterozygotes. Consequently, they usually remain for many generations in the gene pool until a chance mating pairs them to another mutant. Although not impossible, it would be quite unusual for them to occur homozygously and be expressed in the first few generations after they occur.

If the mutant trait is neither better nor less fit for the environment than the nonmutant trait, it will persist in the pool. If it is better fit, it will become successful, and if it is less fit, it will begin to be eliminated. Elimination, however, will be slow. When two heterozygotes carrying the mutation mate and conceive, only one-fourth of their offspring will be homozygote mutants; half will still be heterozygotes (Fig. 3.2). Though the mutation may be disadvantageous, and those who express it may be eliminated, it remains in the gene pool, still masked in the heterozygote by the dominant, nonmutant partner. The number of mutant alleles in the pool will begin to decline, but the allele will not be quickly eliminated. In this manner, even deleterious mutations remain a long time in the pool, dribbling out slowly.

Changes in mutation elimination rate occur, then, not at the time a mutation is added, but usually many generations later when it chances to combine with another recessive like itself and can express its deleterious, life-shortening nature.

The addition of mutations disturbs genetic equilibrium and increases the size of the mutation burden. The burden increases because a mutation (or mutations) has been added which will not be immediately eliminated.

In nature it seems that increases in mutation frequency usually are offset by declines (as when sunspots flare and recede). Presumably the elimination rate also increases and declines—albeit out of phase

with changes in entry rate—and the burden remains, on the whole, constant.

Continuous increase in mutation frequency is, however, another matter. Because of the recessive nature of deleterious mutations, if their frequency were continuously increased, their extinction would lag behind such increases and never be able to catch up. The equilibrium between entry of mutations into a gene pool and elimination of deleterious mutations from the pool would be destroyed. The mutation burden or load would no longer remain constant but would grow, resulting in an *ever-increasing* burden of deleterious mutations in the pool, which would be expressed as genetic disease or disability. Translated to less abstract terms, this means that, under such circumstances, populations of the future would have ever-increasing numbers of heritable diseases or weaknesses which would shorten the *average* life expectancy of their members; in short, continuously increasing mutation burden is expected to produce personal injuries to ever-increasing numbers of persons not yet born. If increases were to continue indefinitely, it is possible that the capacity of the population to survive might be threatened.

Medicine is the principal source of exposure of the human population to man-made radiations, leaving the possibility that gonadal exposure occurring during medical procedures is harming the human populations of the future. This is an important question and one that must be closely examined. Does medical exposure today bring a significant risk to humans of tomorrow?

Two things must be determined if this question is to be answered. They are: (1) the *number* of mutations induced by given doses of ionizing radiations, and (2) the *fraction* of the medical radiation dose given yearly that has *genetic* significance. (Only that fraction of medically-given doses of radiation absorbed in *gonads* of people who

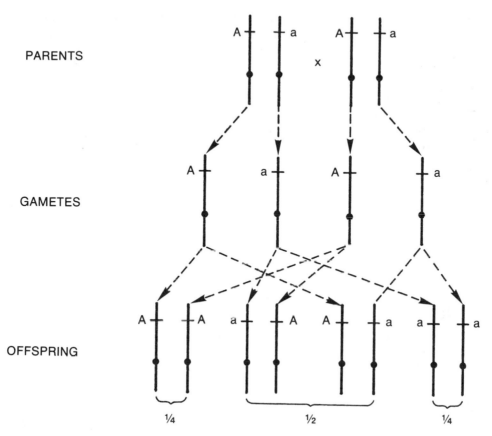

Fig. 3.2. The mating of two heterozygotes may produce offspring with one of three genotypes. One-fourth will be homozygote dominants, one-half heterozygotes, and one-fourth homozygote recessives.

subsequently *reproduce* has genetic impact or significance.)

3.8 GENETICALLY SIGNIFICANT DOSE

Gonads probably absorb some radiation during every medical procedure. When they are in the collimated beam, they receive a relatively high dose, but even when they are outside the beam, they receive radiations scattered from structures *in* the beam. In nuclear medicine procedures they may receive radiation as radionuclides circulate in blood and/or from the various organs which concentrate radionuclides or are involved in their excretion. The *total* medical radiation dose received by a population in any given year and the medical dose that

can have genetic significance will not be the same. The *genetically significant dose* will only be a fraction of the total dose, that fraction absorbed in gonads of persons *who produce children*. Only mutations produced in gametes which are placed in the gene pool (which produce children) are of significance to the population of the future; the remainder die with the individual in whom they were produced.

Because of the serious potential hazard presented by continuously increasing mutation frequency, an effort has been made to determine whether *gonad dose* from medical radiation is on the increase. If gonad dose in persons with childbearing expectancy *is* continuously increasing, it is reasonable to assume that mutation frequency and mutation load may also be increasing.

The genetically significant dose (GSD) from medical procedures, that fraction of medical radiation that can have genetic impact, has been determined at occasional intervals. It is important to keep firmly in mind that the GSD is an index of the *presumed* genetic impact of medical radiation on the whole population. *It cannot be applied to any individual* or to his or her unborn children.

The gene pool consists of all the genes placed in it by any generation. *Both* people who have absorbed some radiation in their gonads from medical procedures *and* people who have not contribute to the pool. Therefore, a situation in which some people put many mutations in the pool and others few has the same effect on the population or pool as a situation in which *everyone* contributing to the pool puts in an equal number. As a hypothetic example, suppose there were a gene pool consisting of 100 members. If 10 members of the pool contribute 10 mutations to the pool, and the remaining 90 contribute none, 100 mutations will have been placed in the pool. The *impact* in the pool, viz., the addition of 100 mutations, is the same as if each member (100 total) had transmitted one mutation each to the pool.

This concept is used in determining the GSD. The estimated gonadal *dose* from medical radiation of those who actually contributed, or had expectation of contributing, to the pool (childbearing expectancy) is adjusted for the size of the pool. For this reason the GSD cannot be applied to individuals; it is devised for and applied to populations only.

3.9   DETERMINATION OF THE GENETICALLY SIGNIFICANT DOSE

In determining the GSD of medical radiations, it is necessary to determine (1) the absorbed dose in the gonads from medical procedures and (2) the probability of childbearing in persons with irradiated gonads.

Gonad Dose

Gonad *exposure* is considered as the sum of two components—photons interacting directly in the gonads without first being scattered in the body (direct radiation) and photons interacting in the gonads after having been scattered from other parts of the body (scattered radiation). When gonads are outside the collimated beam in any procedure, there is no contribution of direct radiation to gonad exposure (see Table 1.2). However, scattered radiation may reach the gonads whether the gonads are within the collimated beam or not. Total gonad exposure (the sum of the direct and scattered components) varies according to the anthropometric characteristics of the irradiated person (height, weight/height quotient, location of gonads, and attenuation distance in tissue).

Childbearing Expectancy

Presumed genetic injury is associated only with *offspring* of irradiated individuals. Therefore, in determining GSD, gonad dose of irradiated persons must be weighted for the probability of production of offspring. Gonadal irradiation of persons beyond the usual childbearing age contributes practically nothing to GSD. Moreover, even during childbearing years, age at irradiation is significant. The *number* of offspring expected from persons who are in the late childbearing years is fewer than from persons who are in the early childbearing years. The future childbearing expectancy for any given year of age is taken from age-specific birth rates *for that year*. Because birth rates change from time to time and even from year to year, childbearing expectancy for persons of any given age can be different in different years. GSD is a function not only of specific gonadal dose but also of the numbers of persons of specific childbearing expectancy irradiated, and it is a variable function of the year for which it is determined.

All of the foregoing, gonad dose given

by specific procedures, number of persons in any given year receiving that gonad dose from medical sources, and childbearing expectancy of those irradiated persons, can only be approximated. As in most such studies, a sample, presumed to be representative of the population as a whole, is carefully studied, and the results are extrapolated to include the whole population. This is the best that can be done short of the Augean labor of studying the entire population. The accuracy of the extrapolation hinges, obviously, on the degree to which the sample represents the general population. The study assumes also that irradiated persons have the same childbearing expectancy as has the general population; but this, in fact, may not be true. Medical conditions necessitating exposure to radiation may affect childbearing expectancy. There is a lack of data on this point; consequently, no adjustment can be made for it.

While the factors pointed out above are sources of inaccuracy in determining GSD, we have not drawn attention to them for the purpose of devaluing the studies leading to GSD determinations. We wish only to indicate the difficulties confronting any such determinations and to caution that any figures obtained for GSD are estimates, albeit the best available.

3.10 GSD FOR 1964 AND 1970

Comparisons of GSD for medical exposure for the years 1964 and 1970 are instructive. The GSD is estimated at 17 millirads in 1964 and 20 millirads in 1970. The difference between them was not regarded as statistically significant ($p = 0.10$). As expected, examinations involving the abdomen result in relatively high gonad doses while those involving head, neck, and thorax result in relatively much lower doses. *Mean* gonad doses for various procedures were used to compute the GSD, but it was noted that maximum and min-

imum gonad doses for given procedures could be several orders of magnitude different from the mean.

Eight procedures account for 90% of the GSD. They are: barium enema; intravenous or retrograde pyelogram; abdomen KUB; flat plate; lumbar spine; pelvis; hip; lower extremities and other abdominal procedures. However, the degree to which these contribute to the GSD differs depending on the sex of the individual examined (Table 3.1).

3.11 GSD AND MUTATION FREQUENCY

Although the GSD would have to be determined for any given year to speak about its effect on mutation frequency for that year, the data of 1964 and 1970 permit some tentative generalizations. For one thing, although the number of persons irradiated and the number of procedures per person increased, there was no *significant* increase in GSD between the years 1964 and 1970. It was estimated that in the United States 66,086,000 persons received 104,987,000 x-ray examinations in 1964, and 76,449,000 persons received 129,070,000 x-ray examinations in 1970. The GSD appeared not to change. This suggests that increasing use of x rays for diagnostic purposes does not necessarily portend increased genetic dam-

Table 3.1. *Contributions to the genetically significant dose (GSD) from eight radiographic examinations (in millirads), 1970.*

Examination	Male	Female	Total
Lumbar spine	0.5	3.1	3.6
Other abdominal	0.8	3.0	3.8
IV pyelograms	1.0	2.2	3.2
Barium enema	0.3	1.7	2.0
Pelvis	0.7	1.1	1.8
Abdomen KUB	0.4	1.1	1.5
Hip	0.7	0.2	0.9
Lower extremity	0.6	0.05	0.6
TOTAL			17.4*

*The total GSD of 20 millirads includes 1 millirad to the fetus, essentially all from these 8 procedures.

age to the population. It further suggests that with good dose-limiting procedures such as collimation, intensification of image, image storing, and others, the GSD can be kept low. It is well to point out here that the GSD of 20 mrem per year for diagnostic x-ray examinations and 2 to 4 mrem per year for radiopharmaceuticals is small compared to exposure from the natural background, which delivers a GSD of between 80 and 100 mrem per year.

The lack of apparent increase in GSD in the face of increased diagnostic use of radiation is encouraging. It suggests that exposure to increasing quantities of medical radiation does not necessarily have to result in continuous increase in mutation frequency. In turn this suggests that, although the amount of genetically significant radiation has undoubtedly increased since medical irradiation was introduced shortly after Roentgen's discovery of x rays, it does not have to increase *continuously*. It is possible that medical gonadal irradiation over the years may have caused an increase in mutation burden with results similar to those produced when there is a period of intensified solar or celestial irradiation, but the rate of increase may have slowed or even reached zero. Medical radiation does not have to present an inordinate risk to the gene pool, and in cases in which irradiation has clear, potential benefit and is carried out in a manner that protects the gonads, it can be more beneficial than hazardous.

While the GSD measures *genetic impact* of medical irradiation, it does not measure or estimate the actual increase in mutation frequency as a result of the GSD. The GSDs that have been determined are small and, presumably, increase mutation frequency only a small amount, but the actual increases are unknown.

3.12 MUTATIONS PER REM

The precise number of mutations produced per rem or rad of ionizing radiations in human beings is not known with certainty. Estimates for humans based principally on lower animal data suggest, however, that it is in the range of about 10^{-7} per gene per rem or rad. If this is so, a GSD of about 17 or 20 mrem will result in an extremely small mutation induction frequency—a minute fraction of the spontaneous rate. However, it is true that the actual increase in mutation frequency from such GSDs remains uncertain because there are a number of unanswered or unsatisfactorily answered questions about the relationship between radiation dose and mutation frequency. In 1974 an ad hoc committee of the National Academy of Sciences[4] reported that a wide variety of biologic end points, including mutation, studied over an extensive dose range, showed a linear-quadratic relationship (Fig. 3.3).

The linear increase in mutation frequency is believed to be due to genetic changes brought about by the action on DNA of single ionizing particles, but genetic changes in the quadratic part of the curve are believed to be produced by the action of more than one ionizing particle.

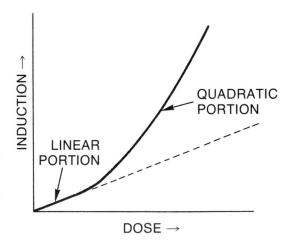

Fig. 3.3. The general relationship believed to exist between mutation frequency and radiation dose. At low dose ranges, mutation frequency is a linear function of dose. At higher doses, the relationship becomes quadratic, i.e., mutation frequency varies as the square of the dose.

In the quadratic part of the curve, lesions produced by individual particles somehow interact, and increase in mutation frequency, *per unit of radiation dose*, is greater than at lower doses.

The linear-quadratic model has not received universal acceptance, but the controversy about it seems to focus mainly on the quadratic.[1,4-7] There is more agreement that the lower range of doses produces either a truly linear increase in mutation frequency or a close approximation of a linear dose-response relationship. However, in the lowest dose range, data are equivocal, and it is impossible to define the relationship with certainty. The doses that are important in medicine, even those used in radiation therapy, produce gonadal doses which, on the whole, fall at this low end of the scale. For example, gonadal doses from diagnostic procedures are principally from scattered exposure; they probably rarely exceed one rad and are often much lower.

The uncertainty over the relationship between radiation dose and mutation-frequency in very low dose ranges is the result of formidable technical problems involved in obtaining unequivocal data. This is especially true for mammals, presumed to be the most relevant model for man. Radiation produces *no* unique mutations; it produces the same ones that occur spontaneously, simply increasing their frequency. Often no *statistically significant* increase in mutation frequency can be shown after exposure of experimental systems to low doses. It is hard to decide whether this means that no increase occurred or that whatever increase may have occurred is too small to be distinguished from the background of spontaneously occurring mutations. Very large populations must be used to detect significant levels of change, and often a large number of generations must be observed as well. Such experiments are time-consuming and frequently prohibitively costly. Yet, to predict precisely how much mutation frequency is increased by

the GSD, the shape of the curve in this low dose range must be known.

The key point, however, from the point of view of the population, is not necessarily how much mutation frequency is increased by the genetically significant dose, but that the GSD be low and not *continuously* increasing. GSD measurements have given some assurance on these points.

3.13 DOUBLING DOSE

With respect to the number of radiation-induced mutations, the frequency increase produced *per rem* or *rad* is estimated to be approximately 10^{-7} per gene for human beings. The spontaneous mutation rate in humans is estimated to be about 10^{-5}/gene/generation. Gonadal exposure of one rem must then increase mutation frequency an extremely small fraction of the spontaneous mutation frequency.

The rate of induced mutation can be stated in absolute terms (the probability of mutation per locus per rad, approximately 10^{-7}), or it can be stated in relative terms, such as the ratio of the induced mutation rate to the spontaneous rate. This ratio, the fraction by which each rad of added exposure would *increase* the mutation rate above the spontaneous level, is called the "relative mutation risk." Given the estimate of 10^{-7}/gene/rad and the spontaneous rate of 10^{-5} per gene, the relative mutation risk would be $10^{-7}/10^{-5}$ which is 1/100 or 1%. Thus, one rad would increase the mutation rate by 1%, a relative risk of 0.01. If the incidence of some specific condition in the population before radiation was 100 per million liveborn, the expected *increase* at equilibrium (perhaps 10 generations later) would be 1 per million. Frequently, the reciprocal of the relative mutation risk, the "doubling" dose, is given. It is the dose required to induce a mutation frequency equal to the spontaneous mutation frequency; hence, the term doubling dose. For instance, a relative mutation rate of 0.01

per rad, a risk of 1/100, would give a doubling dose of 100 rads, a value equal to the estimate given in UNSCEAR, 1977.[24] The 1980 BEIR Report adopted a range of relative mutation risk of 0.004 to 0.02 per rad which gave a doubling dose of 50 to 250 rads,[1] which is in good agreement with the UNSCEAR, 1977, estimate.

A recent publication, basing its conclusions primarily on analysis of the survivors of the atomic explosions at Hiroshima and Nagasaki, suggests that the genetic doubling dose may be toward the upper end of that scale and fixed it at 156 rem for acute exposures.[25]

3.14 MUTATION AND DOSE RATE

When the rate at which radiation is delivered is altered, differences in mutation frequency are observed.[8-11]

Although increase in mutation frequency is affected to a degree by the stage of germ cell maturation in which irradiation occurs,[11] generally it appears that the lower the rate at which radiation is delivered, the less the increase in mutation frequency *for any given dose.* If this phenomenon holds true for mutation induction in humans, it suggests that doubling dose for protracted exposures, such as those encountered occupationally or from the background, might be higher than for acute exposures. Authors reexamining the Hiroshima-Nagasaki experience suggest the chronic doubling dose might be 468 rem, three times their acute estimate.[25]

Dependence of mutation induction on dose rates has led some authors to postulate for the induction of mutations a mechanism that includes a reversible or reparable process. The hypothesis holds that mutation is at least a two-step process and that all steps in the process must occur before a true mutation exists. The first of these steps ("premutational" change or damage) is supposed to be reversible or reparable. If more steps follow premutational change

before premutational change is either repaired or reversed, then a mutation will occur. During irradiation at low dose rates, significant numbers of premutational changes might occur but might also be repaired or reversed before the subsequent events occur that produce the full mutation. No mutation will occur. Increasing dose rate will increase the probability that all events leading to mutation will take place so closely together that repair or reversal of premutational change will not have time to occur. Larger numbers of mutations, *per unit of dose,* will therefore be produced at high rather than at low dose rates.

This hypothesis would be contradicted if a comparison between the changes in mutation frequency when given doses are delivered at one time, and when they are split into two fractions, did not show differences. Comparisons can be made as follows. A particular dose is given to the gonads of test organisms over a short period of time. Then, the same dose is given to the gonads of identical organisms, but the dose is split into two fractions with a time lapse between. Some authors, attempting split-dose experiments, believe their data *support* a recovery or repair hypothesis; mutation frequency was increased *less* when dose was split than when given all at once. Presumably, in the time between delivery of the fractions some premutational changes were repaired or reversed, yielding fewer total mutations.

Another line of evidence which can be taken to support a two- or multi-step mutational process and a recovery or repair hypothesis[12-15] is the change in increase of mutation frequency from given doses of radiations varying in LET. If an analysis is made of mutational changes produced *per viable cell* as a function of LET, radiations of higher LET produce relatively more mutagenic lesions than those of lower LET.[16,17] It is worth mentioning again that high-LET radiations are more lethal than are low-LET radiations. Consequently, the total number of mutations may be greater in a cell population irradiated with low- rather than high-LET radiations. Since more cells *survive* low-LET irradiation, there will be more living cells to express their mutations. However, when one is scoring number of mutations per *viable* cell following low- and high-LET irradiation, the number of mutations is greater following high-LET irradiation.

Increase in mutation production as a function of increasing LET can be taken as evidence that mutation is at least a two-step process. Radiations of low-LET have a smaller expectation of bringing about several changes within a given molecule than do high-LET radiations. If mutation depends on the production of several injurious changes in genes, then high-LET radiation, which may enter into several interactions in small volumes, should be more effective than low-LET radiations in bringing about the effect. Low-LET radiations may produce a relatively large number of *premutational* changes in most nuclei—when given in low doses. If these changes are reversible, these low doses will produce few mutations. At higher doses, the *number* of low-LET radiations to which cell nuclei are exposed is great, and the probability that several ionizations will occur in a given molecule is high. At high doses, low-LET radiations have a good chance of producing premutational as well as any additional events that must take place for mutation to occur.

3.15 LOSS OF MUTATIONS WITH TIME

The probability of inducing mutation in the germ line is a function of the stage of maturation of sperm or ova. In males, germ cells in the late stages of maturation (*following* meiosis)—namely spermatids and spermatozoa—have about twice the probability of mutation induction as do earlier, less mature forms,[1,18] but they are quite resistant to being killed by ionizing radiation.

The chances of males transmitting mutations is highest directly after gonadal irradiation, because the mature forms (sperm and immediate precursors) are radioresistant but highly susceptible to mutation induction.

On the other hand, the immature, premeiotic forms (spermatogonia) are extremely radiosensitive, but those surviving radiation are less susceptible to mutation induction than mature forms. Consequently, for several weeks after testicular irradiation, sperm count remains high (sperm and spermatids are radioresistant) and the probability of fertilization by irradiated sperm appears to be the same as that by unirradiated sperm.[19]

If either patient or physician is concerned about transmission of mutations into the gene pool, good advice would be to suggest that irradiated men wait about three months after testicular irradiation before attempting conception. Sperm appearing then are produced from immature premeiotic forms which have survived irradiation and will have a *relatively* low mutation content. If dose was high enough to reduce sperm count[19] (even 20 rad in men reduces sperm count), the count will probably remain low for up to several years (a phenomenon which, in itself, tends to reduce the chance that mutations will be put into the pool).

Similar advice is *probably* also valid for women whose gonads have been exposed. Data gathered in mice[1,18,19] show that no mutations are transmitted when conception is delayed two months or longer following irradiation. There are no data taken directly from irradiated women, however, and no one is sure whether these mouse data can be extrapolated to humans.

3.16 HIROSHIMA, NAGASAKI AND GENETIC EFFECTS

Children whose parents were irradiated in the atomic bombings of Hiroshima and

Nagasaki have been closely studied, and the studies have been reviewed several times.[20–22,25]

The earlier studies[20–22] showed no differences in any of the measures of health or survival between these children and those of unirradiated Japanese in spite of the fact that many of their parents had received high gonadal doses. The recent reappraisal[25] came to similar but not exactly the same conclusions. It presented data on four indicators of genetic effects: frequency of untoward pregnancy outcomes (still births, major congenital defects, death during the first postnatal week); occurrence of death in liveborn children through an average life expectancy of 17 years; frequency of children with sex chromosome aneuploidy; and frequency of children with mutation resulting in an electrophoretic variant. In none of these did they find a statistically significant effect of parental exposure, but for all of them they observed effects in the direction expected if genetic damage resulted from the exposure. These results, combined with the results of the enormous body of data concerning the genetic effect of radiation on experimental organisms, have led these authors to conclude that genetic damage in fact occurred.

Some investigators have suggested that recessive genetic damage done by parental exposure to the atomic bombs might result in a shift in the sex ratio. Animal experiments have shown a shift in sex ratio when *recessive*, X-chromosome-linked, deleterious or lethal mutations are produced. As described earlier in this chapter, the expression of recessive mutations is usually masked by the presence of dominant, nonmutant alleles on homologous chromosomes. In females, there are two X chromosomes which are homologs of each other, and recessive mutations produced on one will be masked by the dominant, nonmutant allele on the other. In males there is only one X chromosome. The Y chromosome frequently has no allele for genes on the X chromosome. Recessive

mutations produced on the X are, therefore, not always dominated by an allele on the Y and are expressed. If they are deleterious, and especially if they are lethal, they cause a shift in sex ratio.

Generally, in humans, males and females are born in about equal numbers. In populations in which there are severely deleterious or lethal X-linked recessive mutations, males bearing them die (often at or before birth), because there is no dominant allele on the Y chromosome to mask them. Females bearing these same severely deleterious recessives live, because there may be a dominant allele on the other X chromosome that does mask their expression. Such populations do not have a 50:50 sex ratio but one in which there has been a shift—more females being born than males.

If many severely deleterious or lethal mutations were produced at Hiroshima and Nagasaki, some inevitably would have occurred on the X chromosome, and a sex ratio shift in favor of females would have occurred. Numerous studies seeking such a shift have been made and most detected none.[23] Some results gathered early after the explosions showed a barely significant shift in the expected direction, but later results failed to confirm it.

The lack of sex ratio shift in the first generation should not be interpreted as meaning that *no* recessive mutations were induced at Hiroshima and Nagasaki. Possibly, studies of future generations will reveal the existence of mutations not expressed in the first generation. All that may be concluded is that no such effects have *as yet* been seen. Moreover, the recent reappraisal[25] suggests that sex ratio is a more complex indicator than just outlined and that genetic damage may occur without necessarily causing a perceptible shift.

3.17 ESTIMATED RISK IN HUMANS IN THE FIRST GENERATION

Often the concern of patients and occupationally exposed people is the potential genetic effect on their immediate descendants, children and/or grandchildren. Induced recessive mutations are rarely expressed so soon, but occasionally mutations are induced that are dominant. These are believed responsible for most serious handicaps observed in the first few postirradiation generations.

Most data on dominant disorders are taken from skeletal abnormalities occurring in the first descendant generation of male mice, and this information is used to estimate the number of dominant mutations expected in humans per unit of gonadal radiation dose.

The BEIR[1] report states that 107,000 out of one million live births (about 10%) carry a serious spontaneously induced genetic disorder. It estimates that parental gonadal exposure of *one rem* would result in an additional 5 to 75 cases in the first generation. These numbers illustrate the great difficulty in measuring directly such small increases against so large a background of spontaneous mutations.

Estimates for the risk in the first postirradiation generation and for the risk at equilibrium from 1 rem parental exposure per generation over many generations are given in Table 3.2.

3.18 CONCLUSIONS

The foregoing strongly suggests that mutation induction resulting in gene pool damage need not be the *primary* concern of those using, receiving, or working with medical radiation. While irradiation indisputably increases mutation frequency, it appears that, at low doses, the increase per rem or rad is an extremely small fraction of the spontaneous mutation frequency (Section 3.12). Moreover, as the genetically significant dose (GSD) appears to be a small fraction of a rem (Section 3.11), exposure for medical reasons probably adds a very small amount, indeed, to the burden of the gene pool.

The increase in mutation frequency in individuals irradiated during diagnostic procedures is also likely to be miniscule. The doubling dose (Section 3.13) is large compared to average gonadal doses (see Table 1.3), so the risk of transmitting large numbers of mutations as a result of diagnostic irradiation is almost insignificant. Furthermore, it can be argued that some DNA damage is reversible or reparable and does not lead inevitably to mutation (Section 3.14), a phenomenon which may be especially important after very low gonadal doses.

The risk of mutation may be minimized by delaying conception several months after gonadal exposure (Section 3.15). Mutation induction rate is smaller in premeiotic gametic stages than in later stages of germ cell maturation. Also, premeiotic stages are radiosensitive. Doses in the range of the doubling dose (which can occur during radiotherapy) reduce fertility and the likelihood of transmitting mutations as well (Section 3.15).

So far as radiation workers are concerned, the maximum gonadal dose permitted by law is five rem per year. Clearly, if the various calculations are correct, this should increase the mutation frequency by a tiny, if not insignificant, fraction of the spontaneous rate.

The absence of significant effects in physical measurements, health, or survival in children of survivors of the high doses received at Hiroshima and Nagasaki are also hopeful signs (Section 3.16).

The fact that there are so many signs pointing to a very small effect on the gene pool from medical and occupational exposure is no license to be cavalier about gonadal irradiation. For the present hopeful situation to continue, strict attention must be paid to keeping gonadal doses to the practical minimum. Tampering with the gene pool is a serious matter and should be regarded as such by every user of radiation. The benefits of irradiation may be enjoyed by patient and worker, but the risks are borne by other people, the population of the future. We must be wary on their account.

3.19 REDUCTION OF GENETIC RISK

The UNSCEAR Report[24] recommends the following ways to reduce gonadal dose.
1. Reduce the number of radiographs per patient.
2. Reduce the time and intensity of exposure.
3. Use radiographs only, when fluoroscopy is not essential.
4. Use the smallest possible field size.
5. Avoid inclusion of the gonads in the primary beam.
6. Protect testicles with gonadal shields.
7. Properly train and supervise staff engaged in these examinations.

During medical exposures gonads should be directly exposed to the collimated beam of radiations only when absolutely unavoidable. Otherwise all measures should be taken to protect them against exposure so that gonadal dose is minimized. While portions of the body in direct, collimated beams of radiation receive the full dose given by particular procedures (diagnostic or therapeutic), gonads should receive only the radiations scattered to them from the region of the body directly exposed. Gonadal dose should generally be only a fraction of doses absorbed by tissues directly exposed.

SUMMARY

1. Point mutations occur spontaneously in genes.
2. Many point mutations are recessive and deleterious.
3. Exposure to ionizing radiation increases mutation frequency. Estimates are that approximately 10^{-7} mutations are produced per rem per

gene, an extremely small fraction of the spontaneous mutation rate (about 10^{-5}/gene/generation).
4. The doubling dose for mutations is estimated to be about 50 to 250 rem, a recent report giving it as 156 rem.
5. The genetically significant dose, that radiation dose having genetic impact, was 17 and 20 millirad for 1964 and 1970 respectively, numbers *not* significantly different from each other.
6. The genetically significant dose does not appear to be increasing and presumably raised the human mutation frequency less than 10^{-7} per gene (see 3 preceding).
7. The relationship between mutation frequency and radiation dose is linear-quadratic for low-LET radiations.
8. Mutation frequency changes are influenced by LET and dose rate.
9. More mutations are induced in postmeiotic germ cells than in premeiotic germ cells. The probability of transmission of mutation is greatest within

Table 3.2. *Summary of low-dose risks*

GENETIC DEFECTS

First Generation

Number Affected per Million Liveborn from 1 rem Parental Radiation[a]	Source
5–75 20–30	BEIR, 1980 UNSCEAR, 1977

At equilibrium

Number Affected per Million Liveborn from 1 rem Parental Radiation over Many Generations[a]	Source
60–1100 185	BEIR, 1980 UNSCEAR, 1977

[a]Spontaneous incidence = 107,000/million liveborn

the first two to three months following gonadal irradiation, when mature forms are present, and declines thereafter (to one-half in men, unknown in women) as new gametes mature.

10. Premeiotic germ cells develop fewer mutations per unit of radiation dose than do post-meiotic forms and the premeiotic cells are exquisitely radiosensitive. Doses of 20 rad (at the lower end of doubling dose estimates) may induce long-term inhibition of fertility.

11. No differences in measures of health or survival in offspring of survivors of atomic bomb explosions at Hiroshima and Nagasaki compared to other Japanese children were noted. If serious genetic damage was done, no evidence of it appeared in the first generation.

12. Table 3.2 summarizes estimated risk of genetic defects following gonadal exposure to low-dose radiation.

REFERENCES

1. National Research Council, Advisory Committee on the Biological Effects of Ionizing Radiation (BEIR III): The Effects on Populations of Exposure to Low Levels of Ionizing Radiation. Washington, D.C., National Research Council, 1980.
2. United Nations Scientific Committee on the Effects of Atomic Radiation (UNSCEAR): Ionizing Radiation: Levels and Effects. New York, United Nations, 1977.
3. Levitan, M., and Montagu, A.: Textbook of Human Genetics. New York, Oxford University Press, 1971, p. 682.
4. National Academy of Sciences: Research Needs for Estimating the Biological Hazards of Low Doses of Ionizing Radiation. Washington, D.C., 1974.
5. Kellerer, A.M., and Rossi, H.H.: The theory of dual radiations action. Curr. Top. Radiat. Res. Q., 8:85–158, 1972.
6. Abrahamson, S.: Mutation process at low or high radiation doses. In Biological and Environmental Effects of Low Level Radiation. Vol. I. Vienna, International Atomic Energy Agency, 1976, pp. 3–7.
7. Abrahamson, S., and Meyer, H.U.: Quadratic analysis for induction of recessive lethal mutations in Drosophila oogonia by x irradiation. In Biological and Environmental Effects of Low Level Radiation. Vol. I. Vienna, IAEA, 1976, pp. 9–17.
8. Russell, W.L.: Evidence from mice concerning the nature of the mutation process. In Genetics Today. Proc. Internat. Cong. Genet. XI. Vol. 2. Oxford, Pergamon Press, 1964, p. 257.
9. Sobels, F.H.: Chemical steps involved in the production of mutations and chromosome aberrations by x-irradiation in Drosophila. I. The effect of post-treatment with cyanide in relation to dose-rate and oxygen tension. Int. J. Radiat. Biol., 2:68–90, 1960.
10. Sobels, F.H.: Genetic repair phenomena and dose-rate effects in animals. In Advances in Biological and Medical Physics. Vol. 12. Edited by J. Lawrence and J. Gofman. New York, Academic Press, 1968.
11. Russell, W.L.: The different effects of dose-rate on radiation-induced mutation frequency in various germ cell stages of the mouse and their implications for the analysis of tumorigenesis. In Radiation Biology in Cancer Research. Edited by R. E. Meyer and M.R. Withers. New York, Raven Press, 1980, pp. 321–325.
12. Muller, H.J.: Radiation genetics. In Radiation Biology. Edited by A.J. Hollander. New York, McGraw-Hill, 1954, p. 351.
13. Fox, M.: Repair synthesis and induction of thymidine-resistant variants in mouse lymphoma cells of different radiosensitivity. Mutat. Res., 23:129–156, 1974.
14. Azlett, C.F., and Potter, J.: Mutation to 8-aza-buanine resistance induced by x-radiation in a Chinese hamster cell line. Mutat. Res., 13:59–65, 1971.
15. Deem, D.F., and Shaw, E.I.: Recovery during radiation mutagenesis In Biological and Environmental Effects of Low-Level Radiation. Vol. I. Vienna, IAEA, 1976, pp. 19–29.
16. Cox, R., Thacker, J., Goodhead, D.T., and Munson, R.J.: Mutations and inactivation of mammalian cells by various ionizing radiations. Nature, 267:425–427, 1977.
17. Goodhead, D.T., Munson, R.J., Thacker, J., and Cox, R.: Mutations and inactivation of cultured mammalian cells exposed to beams of accelerated heavy ions. IV Biophysical interpretation. Int. J. Radiat. Biol., 37:135–167, 1979.
18. Searle, A.G.: Mutation induction in mice. Adv. Radiat. Biol., 4:131–207, 1974.
19. Rowley, M.J., Leach, D.R., Warner, G.A., and Heller, C.G.: Effects of graded doses of ionizing radiation on the human testis. Radiat. Res., 59:665–678, 1974.
20. Neel, J.V., and Shull, W.J.: The Effect of Exposure to the Atomic Bombs on Pregnancy Termination in Hiroshima and Nagasaki. National Academy of Sciences Publication 461, New York, 1956.
21. Kato, H., Shull, W.J., and Neel, J.V.: Survival in children of parents exposed to the atomic bomb, a cohort-type study. Am. J. Hum. Genet., 18:339–373, 1966.
22. Shull, W.J., Neel, J.V., and Hashizume, A.: Some further observations on the sex ratio among infants born to survivors of the atomic bombings of

Hiroshima and Nagasaki. Am. J. Hum. Genet., *18*:328–338, 1966.

23. Jablon, S., and Kato, H.: Sex ratio in offspring of survivors exposed prenatally to the atomic bombs in Hiroshima and Nagasaki. Am. J. Epidemiol. *93*:253–258, 1971.

24. United Nations Scientific Committee on the Effects of Atomic Radiation (UNSCEAR): Report to the General Assembly, Supplement 100.16 (A/5216), New York, United Nations, 1977.

25. Schull, W.J., Otake, M., and Neel, J.V.: Genetic effects of the atomic bombs: A reappraisal. Science, *213*:1220–1227, 1981.

Chapter 4

Mutation—The Effects of Irradiation on the Chromosomes

4.1 INTRODUCTION

Function of genetic material can be altered not only by point mutations but also by changes in the structure of chromosomes. Chromosomes, bodies found in nuclei of cells, are composed of DNA and protein. Two of their important functions are (1) to preserve the order or sequence of genes, and (2) to assure proper distribution or dosages of genetic material to daughter cells at cell division.

Genes are arranged on chromosomes in a linear order, like pearls on a string, so that there is a fixed spatial relationship among genes on given chromosomes. In the past most geneticists believed that each gene acted as an independent entity, and that its expression was independent of its location. There is now evidence suggesting that the expression of genes can be modified if a change occurs in their linear order.[1,2] A change in the *order* of genes on chromosomes means certain genes will be in a new location, adjacent to genes different from those that flanked it before. The change in location apparently may affect gene expression or timing of gene action. Changes in expression or timing attributed to positional changes can be very significant. For example, some believe that differences between organisms in certain taxonomic divisions may be due, not to different genes, but to differences in position of the same genes on chromosomes. The rather considerable and obvious differences between man and chimpanzee, easily observed at the organism level, is hard to account for by differences in genes between the two organisms. Analysis of the proteins of man and chimpanzee shows more than 99% of them to be identical.[3] The genes of these organisms must be very much alike. Yet man and chimpanzee differ so markedly in brain size; anatomy of the pelvis, foot, and jaws; posture; mode of locomotion; methods of procuring food; and means of communication (in a phrase— anatomy and way of life) that they are usually placed in separate taxonomic families.[4,7] The genetic distance between man and chimpanzee, as reflected by the striking similarity of their proteins, is extremely small, well within the range of much more closely related groups. But the *actual distance* appears much greater.

One of the things which does differ between man and chimpanzee and which may account for the obvious differences between them is the *order* of genes on their chromosomes. Studies of chromosomes of these organisms suggest that, while most of the genes or DNA molecules are the same, the *positions* they occupy on the chromosomes are not.[8,9]

Such observations are not limited to man and ape, but have been made in *Drosophila*,[10] birds, frogs and certain mammals.[11-13]

The order of genes on chromosomes then, in the opinion of a number of investigators, is important. The expression or timing of action of genes may depend on it, and the differences which may result from position differences can be considerable.

The second function of chromosomes— that of properly distributing genes to daughter cells during cell division—is also important. Cells may die or function abnormally if they receive either too many or too few genes at cell division. For proper cellular function, cells must have a full set of genes (none should be lacking) and the balance among them must also be proper. That means that for given genes there must not be either an excess or shortage compared to the quantity of the remaining genes.

Exposure to ionizing radiation has been shown to disturb both chromosome functions discussed above; it alters chromosomal structure, and it causes genic under- and overdosages.

4.2 CHROMOSOME MORPHOLOGY

Chromosomes are thread-like structures of DNA and protein found in cell nuclei. Most of the chromosome readily takes histologic stain, and it is this property from which its name (colored body) is derived. Somewhere along the length of chromosomes (the location is constant for any given chromosome but varies among different chromosomes) is a region which does not

stain. This is called variously the *centromere* or *kinetocore* and is the region of the chromosome which attaches to the mitotic spindle (Fig. 4.1).

The genes are located on the arms of chromosomes and have the special linear order mentioned earlier. In nature particular genes are always located on the same chromosome, and the distance between genes on any given chromosome is constant. The chromosomal location of some genes is known in man, but in well-studied lower organisms (fruit flies and mice), the location of very nearly every gene is known.

4.3 CHROMOSOME BREAKS

The arms of chromosomes are subject to breakage. No reason for this is known, but in nature, it appears that in a fraction of the cells in any organism, the nuclei have one or more chromosomes in which an arm (or arms) is broken. There is a naturally occurring frequency of breaks, known as the spontaneous frequency, which forms a kind of background of chromosome breaks.

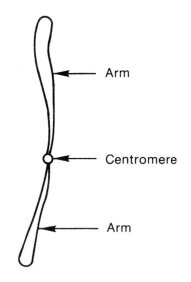

Fig. 4.1. An illustration of a chromosome showing its major parts. Centromeres may be located anywhere along the length of chromosomes, but in normal chromosomes, they are not terminal.

Exposure to radiation increases the *frequency* of breaks above this background.

Chromosome breaks entail a risk for cells in which they occur. When arms are broken the possibility exists that some genetic matter will be lost. This happens, however, *only* if cells with broken chromosomes later divide. If a break (or breaks) occurs in chromosomes in cells which never divide, loss of genetic material does not happen. Despite injury to chromosomes, unless genetic matter is lost, cells seem to function normally. In those in which genetic material is lost, there almost inevitably occurs some loss of function or functions, and very often such cells die.

Breaks in the arms of chromosomes occur when ionizing radiations pass through the chromosomal thread. The thread breaks into parts, but broken ends of chromosomes are sticky, and because of this, frequently the parts stick together again. This process of healing, called restitution, occurs following most chromosome breaks, probably in more than 90% of them. Restituted chromosomes either lose no genetic matter or so little that cells bearing them function normally enough to escape detection. Occasionally, however, restitution does not follow a break. Precise reasons for this are not known, but it seems that if chromosomes are in motion, the probability of restitution is lessened. During cell division, chromosomes are in motion, and most unhealed breaks seem to occur then. The significance of chromosome breaks is that they have a high likelihood of resulting either in loss of genetic material or in rearrangement of genes on chromosomes (Fig. 4.2).

Loss of genes occurs when a chromosome arm is broken, restitution fails to occur, and cell division follows. The detached fragment will have no means of attaching itself to the mitotic spindle, and, consequently, no means of moving with the rest of the chromosomes to the poles of the dividing cell. When division occurs and two daughter cells form, chromosomes should have moved along and been guided by spindle fibers into the nucleus of each daughter. The fragment, lacking an attachment to the spindle, finds this impossible, so it is not included in the nucleus of either daughter. Usually, it lies in the cytoplasm of one of them. Chance apparently determines which daughter receives the fragment. One daughter will have none of the genes on the fragment and the other will have them all; in fact, they will be present in that cell in duplicate, but they will not be in its nucleus. For the cell lacking the fragment there is loss of some quantity of genetic material, the exact quantity determined by the distance from the centromere at which the chromosome was originally broken. In most cases (though not all) loss of chromosome fragments entails loss of sufficient genetic material to impair cell function so severely that the cells die. Cells containing extranuclear duplicated chromosome fragments are not much better off. They may undergo several divisions. Each division again results in one daughter with the fragment and one without. The daughter without the fragment is liable to die because of lack of essential genes, and the one with the fragment builds up ever increasing gene overdosages. This happens because the fragments can never acquire an attachment to the spindle and, duplicated at each division, they accumulate. Ultimately, cells die because of severe genic imbalances. Breakage of chromosomes without restitution is lethal—*provided that cell division occurs after the chromosome break.* Cell death, as a result of chromosome breakage, is known as *mitotic death.* It is a principal mechanism of cell killing in radiotherapy.

Not every chromosome break that fails to restitute is lethal, even if cell division does intervene. It can happen that a break occurs near the tip of a chromosome arm so that genic deficit is small, or a small part of a chromosome may be lost in the region of a break before healing. Cells in which this happens probably have some loss of

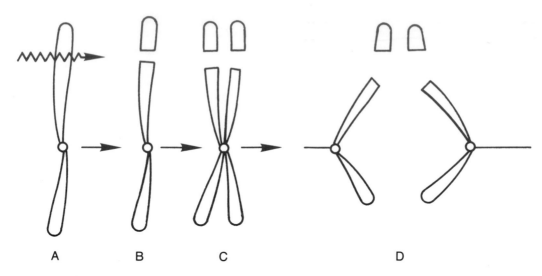

Fig. 4.2. The loss of genetic material. A. A chromosome arm is broken by the passage of an ionizing radiation. B. Restitution does not take place; a chromosomal fragment is created. C. During cell reproduction, the chromosome duplicates (a normal function) and so does the fragment. D. Later in division, spindle fibers attach to the centromeres, and the chromosomes move toward the poles of the cell where, after division, they will become the nuclei of the daughter cells. The duplicated fragment has no attachment to the spindle fiber and does not move. Daughter cells are formed, and the duplicated fragment lies on one side of the cleavage plane in the cytoplasm of one of the daughters.

function, and that functional deficit may be deleterious although not lethal, either to them or to the organism of which they are a part.

4.4 SIMPLE AND COMPLEX CHROMOSOME ABERRATIONS

Chromosome aberrations can conveniently be separated into two categories, simple and complex. A break in one chromosome arm is a simple aberration but *two* or more breaks in the arms of chromosomes in given cell nuclei can result in complex aberrations. While simple aberrations (see Fig. 4.2) cause loss of genetic material, more than one break permits chromosomal rearrangements. Many types of rearrangements are possible. Common ones include translocation (a chromosomal fragment attaches to another chromosome in a nucleus), inversion (a chromosomal fragment rotates before healing), ring chromosomes (the tips of two arms of the same chromosome are broken, and the fragment with

the spindle attachment heals to itself), and dicentrics (two chromosomes are broken, and the fragments containing the centromeres heal together, forming a chromosome having two centromeres (Fig. 4.3).

While these rearrangements are typical of those that occur after irradiation, they are not the only ones. The more breaks that occur in a single nucleus, the greater and more complex are the possibilities for rearrangements.

Rearrangements change gene order. For example, inversions change the gene sequences at the point of healing; translocated chromosome fragments cause a change in "genic neighborhood" at the point of attachment; there are "foreign" genes at the point of healing between inserted fragments and broken chromosomes; at the point of healing of ring chromosomes, a new gene sequence is established; and, of course, there are new gene sequences at the point of healing in the two chromosomes constituting a dicentric. Rearrangements do not necessarily present a barrier to mitosis. Inversions,

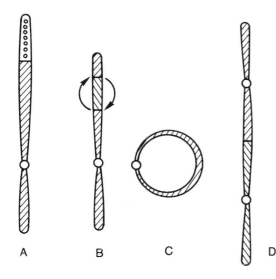

A　　B　　C　　D

Fig. 4.3. Illustrations of complex chromosome rearrangements. A. Translocation—a portion of a chromosome is attached to another chromosome. B. Inversion—a chromosomal fragment rotates before healing. C. Ring chromosome—the broken ends of two arms of the same chromosome have healed. D. Dicentric chromosome—the broken arms of two chromosomes have healed.

translocations, and insertions are examples of this. They may proliferate but, because of *rearranged* genes, these cells and their descendants may act abnormally. In other instances, rearrangement interferes with cell division. Ring chromosomes cannot always pass through mitosis, and dicentrics rarely can. The mechanical difficulty they encounter during complicated mitotic movements causes loss of large parts of or even entire chromosomes and subsequent mitotic death.

Chromosomal rearrangements which do not interfere with mitosis may nevertheless present a barrier to *meiosis,* the cell divisions which lead to the production of gametes. Meiosis is more complicated than mitosis. During mitosis, chromosomes duplicate and then divide, providing daughter cells with exact duplicates of the genetic matter of the parent cell. Meiosis consists of two consecutive divisions. During the first of these, genetic matter is duplicated, and parts of homologous chromosomes are exchanged, shuffling the genetic material.

No duplication of genetic material occurs during the second division, and as a result, the amount passed to daughter cells is halved. Meiois, in contrast to mitosis, does not result in daughter cells having genetic material identical to the parent. Instead, it results in cells having different gene arrangements and half the number of chromosomes of the parent cell.

The essence and purpose of meiosis is to promote genetic variation and assure that genetic rigidity and stagnation do not occur. Shuffling of the genes, so essential to the process, depends first on an intimate pairing (synapsis) between homologous chromosomes. Proper synaptic pairing, in turn, depends on having *identical* gene sequences on homologous chromosomes. Chromosomal rearrangements as a result of irradiation—or any other agent—change the order of genes and interfere with proper synaptic pairing. Shuffling of genes may not occur or may be abnormal because proper pairing was not possible. Finally, rearrangements may cause mechanical difficulties in separation of chromosomes, particularly in the first meiotic division, resulting in loss of genetic matter.

The obstacles chromosomal rearrangements present to gametogenesis may on the whole be a desirable phenomenon. The expression or timing of genic expression may differ between chromosomes with normal genic sequences and those having rearranged sequences (see Section 4.1). Since the precursors to gametes which have rearranged sequences do not always produce gametes, the probability of transmitting genes having abnormal expression or deviant timing is reduced. This should not be taken as meaning that production of chromosome breaks and rearrangements in germ line cells of the gametes is trivial and causes no risk. It is meant only to suggest that natural mechanisms reduce the probability that at least some such rearrangements will enter the gene pool. The natural filtering mechanisms are, however, unlikely to be fully effective, and introduction of rearrangements in the germ line carries some chance that some will be transmitted to future generations.

4.5 CHROMOSOME REARRANGEMENTS IN SOMATIC CELLS

A potential area of great concern is the production of breaks and rearrangements in proliferative, somatic cells. Since rearrangements do not necessarily present obstacles to *mitosis,* proliferative cells containing them may successfully produce a line of descendants. Furthermore, since rearranged gene sequences may not function normally, this is a means of producing within an organism a group of cells containing an abnormally functioning genotype. The degree to which this may affect the organism containing it depends on the particular type of abnormal function and the *number* of cells expressing abnormal function. The former depends on which genes are relocated and how they interact with their neighbors. Little is presently known about this. The latter, the number of cells expressing abnormal function (the number of cells in the clone), depends on the proliferative future of the cell in which the rearrangement first occurred. If the cell is one destined to have many descendants, it will produce a larger clone with presumably greater impact than one destined to have few descendants. Generally, the earlier in life rearrangements occur, the greater their impact is apt to be. For example, cells of an early embryo or fetus are destined to have many descendants. Aberrant chromosomes produced in them could lead to abnormal function over large areas of the adult body. On the contrary, aberrations produced in cells in slowly proliferating adult tissues probably affect very small areas of the body. Quite conceivably, such small areas are affected that abnormal function may frequently go undetected.

4.6 THE INFLUENCE OF DOSE, LET, AND DOSE RATE

The frequency of chromosomal breaks and rearrangements is dependent on dose of radiation. However, the *shape* of dose-response curves and the effect of given quantities of dose are influenced by LET and dose rate of radiations used. Breaks most likely occur as the result of the passage of a single ionizing particle through a chromosome and probably are linear dose functions. Most rearrangements, on the other hand, result from at least two breaks and therefore from the passage of at least two ionizing particles. They occur as a quadratic function of dose.[14] Through low-radiation dose ranges, because there are few radiations available to pass through chromosomes, the radiation effect on chromosomes is mainly the production of single breaks. As dose is increased, however, more and more radiations pass through cellular nuclei, and the probability that some chromosomes will be broken more than once increases. When dose is so high that the probability of several breaks occurring in the same nucleus is very high, rearrangements may begin to occur. If the dose continues to increase, the probability of several breaks occurring in given chromosomes continues to increase, and the yield of rearrangements increases correspondingly. The effect of radiation dose on chromosomes gives a curve with two components, one linear and one quadratic.

Because chromosome breaks are capable of restitution, a *dose-rate* effect occurs in radiation-produced chromosomal aberrations. If the dose rate of radiations is low, a break in a chromosome may restitute before a second break occurs, and this may restitute before a third occurs, and so on. The *number* of unrestituted breaks *per unit of radiation dose* and the number of rearrangements will be smaller after irradiation at low dose rates than at high dose rates, because these generally depend on having *more than one break* open in a given nucleus at a given time.

High-LET radiations are more *effective* at producing chromosome breaks than are those of low LET. An example is found in the observations of chromosome anomalies (exchanges, rearrangements) in lymphocytes of survivors of the atomic bomb attacks on Japan (Fig. 4.4). A strong quadratic term is seen in the Nagasaki survivors who were exposed mainly to low-LET gamma rays. Among survivors of the Hiroshima bomb, which had a neutron component as well as low-LET radiations, data fit a linear dose-response curve, independent of dose rate, which is typical of response to high-LET radiations.[14]

Because energy exchanges with matter are widely spaced with low-LET radiations, much of the track of most particles has only a small probability of releasing enough energy in chromosomes to produce breaks. Near the end of the low-LET track, where the particle has spent most of its energy and is nearly at rest, energy exchanges are probably close together enough to have a high probability of breaking chromosomes. Consequently, only a *portion* of the tracks of low-LET radiations has a good chance of making a chromosome break. Many low-LET radiations have to pass through cell nuclei in order to assure that the tail end of a few of them will pass through and break chromosomes.

The tracks of high-LET radiations are dense throughout, and it is likely that irrespective of what portion of the track passes through a chromosome, the chromosome is liable to breakage.

Although there are not many data, chromosome breaks and aberrations apparently occur in humans after very low radiation doses, such as those in the diagnostic range. It is difficult to make unequivocal statements about this for several reasons. First, as previously mentioned, there is a natural background of chromosome breaks in all populations. Where humans are concerned, it appears that particular individuals may deviate quite considerably from the average background rate. To determine *accurately* increases in frequency caused by small radiation doses, it is necessary to determine first the spontaneous rate for given individuals or groups before irradiation. That has not (and sometimes cannot) be done. Second, the cell type may be important. Certain cell types have a disposition

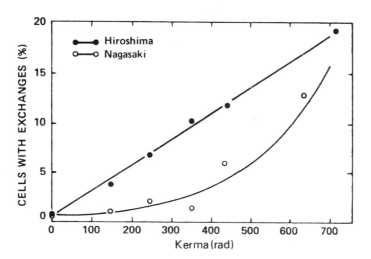

Fig. 4.4. Frequency of cells with exchanges in lymphocyte cultures from Hiroshima and Nagasaki A-bomb survivors 25 years after exposure. (Data from Awa, 1975.) From NCRP.[14]

to breaks and rearrangements, and these are often the types in which analysis is easiest and most often performed. It is not certain, however, that data collected using these cell types can be extrapolated to other cell types in which, because of technical difficulties, data are scarce. Finally, diagnostic radiations are used most often in persons in whom there is disease. Possibly the presence of various disease processes affects fragility of chromosomes, and data collected in such people may not fit everyone. In spite of these cautionary statements, it seems quite likely that diagnostic doses of radiation do increase the frequency of breaks and aberrations in at least certain cells in a portion of the human population.

Increases in aberration frequency have been observed among occupationally exposed persons[15-17] but, so far, no pathology related to these observations has been identified. It is not clear whether there has been no pathology caused by the aberrations or whether the methods so far used have failed to identify it.

4.7 RADIATION EFFECT ON CENTROMERES

Radiations may affect chromosomic behavior if they injure spindle attachment regions, the centromeres. Some common effects are the production of terminal centromeres, chromosome lagging, and nondisjunction. In the case of terminal centromeres, the chromosome is broken in such a manner that centromeres, which in nature are never terminal, now are at the end of the chromosome. For some reason, in that position they do not function normally and, during mitosis, chromosomes with terminal centromeres can be lost.

In the case of chromosome lagging, a chromosome with an injured centromere (not necessarily a terminal one) lags behind the others as they move on the spindle to their respective poles during mitosis.

The lagging chromosome often fails to be included in the newly formed daughter nucleus and remains in the cytoplasm as a micronucleus. This is an abnormal situation, and in time such cells or their descendants die out. The injury to the chromosomal centromere which produces lagging is therefore a lethal injury, but one which is expressed only if mitosis intervenes.

Nondisjunction is the failure of centromeres to separate during mitosis. In the normal course of events, chromosomal arms duplicate and centromeres enlarge during cell reproduction. Spindle fibers attach to the centromeres, and the enlarged centromere splits. Daughter chromosomes are then carried to their respective poles by fibers attached to the centromeres. Irradiated centromeres do not always split after spindle fibers have attached to their sides. The fibers attempt to draw the duplicated chromosome to both poles simultaneously, and in many instances the chromosome, under equal pressure from opposing fibers, does not move at all. Daughter cells form with the chromosome lying between them. It remains there and is not included in either cell, an event which is usually lethal (Fig. 4.5). Occasionally, the pull of one spindle fiber may exceed that of the other so that the entire duplicated chromosome is incorporated in one daughter cell. The other daughter receives no chromosome at all, and genic imbalance is produced in both cells (see Fig. 4.5).

Cells which lack a chromosome are usually not viable, but those containing an extra chromosome may survive and multiply. They have an overdosage of genes; those on their duplicated chromosome are present in excess. This condition, *polysomy*, is deleterious; populations of individuals containing polysomic cells have a shortened *average* life expectancy. They are not as successful as populations with normal chromosome complements and ultimately die out. For instance, as in those with Down's syndrome, a lengthy period of time

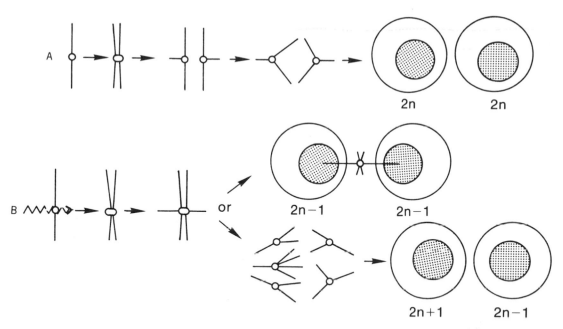

Fig. 4.5. A. Normal mitosis. The centromeres of chromosomes enlarge, and the arms duplicate. The centromere splits, spindle fibers attach, chromosomes are drawn by these fibers in opposite directions and are included in the nuclei of the two new cells that are formed. Each daughter has the 2n number of chromosomes and is an exact duplicate of its parent. B. Nondisjunction. Radiation injures a centromere. It enlarges but cannot split. Spindle fibers attach and attempt to draw the chromosome in opposite directions. The chromosome may remain strung between the two new daughter cells or, if tension on the chromosome is unequal, the duplicated chromosome may be drawn into one of the two daughter nuclei. In the former instance, neither daughter receives the chromosome, and each has 2n − 1 chromosomes. In the latter, one daughter receives an extra chromosome (2n + 1), and the other lacks the same chromosome (2n − 1).

may intrude between the nondisjunction (which occurs either during gamete maturation or during the first or second cleavage division of the embryo) and the death of an organism composed of cells descended from it.

Most malignant cells either lack a chromosome or part of one or have extra parts of whole chromosomes[18] and are nonetheless viable. It is uncertain, however, whether these genic imbalances are the cause or result of malignancy. In any event, they are often associated.

Generally, then, radiation effects on chromosomes are lethal to cells only if they try to reproduce. Cell death probably results from loss of relatively large numbers of genes. This, in turn, results from loss of attachment to the mitotic spindle of chromosomal fragments broken from the chromosome thread by the radiation interac-

tion. Loss of fragments can occur only at or during cell division because it is then that the spindle is formed and attachment to it is important.

Chromosomes can undergo rearrangements which are not necessarily lethal or do not necessarily result in gene loss. Rearrangement of genes on chromosomes evidently can affect gene expression and may produce function different from that normally observed.

Radiation effects on chromosomal centromeres can be lethal either immediately after mitosis or not until several cell generations later.

Finally, the production of chromosome aberrations is affected by the presence or absence of oxygen. Oxygen enhances the ability of low-LET radiations to produce aberrations, but oxygen enhancement is less pronounced if irradiation is carried out with high-LET radiations.

4.8 POSSIBLE RESULTS OF CHROMOSOMAL DAMAGE

If chromosomes are damaged in cells which proliferate, and if the damage does not cause lethality during division (mitotic death), the possible results can be serious and far-reaching. For example, abnormal numbers of chromosomes or abnormal chromosomes are often associated with cancers.[18] It is common for cancers to have extra chromosomes or to be missing whole chromosomes. These deviations from proper chromosome number can occur through lagging or nondisjunction, effects caused by irradiation of the centromere or by other injurious agents. Frequently, cancer cells have portions of chromosomes deleted or parts inverted or translocated to other chromosomes. These aberrations also are caused by irradiation and by other agents.

While chromosome aberrations that can be produced by irradiation are the same as those associated with cancers, no one knows what the association means. It may be that the genic imbalances are somehow involved in transformation from the normal to the malignant state, that malignancy is a result of gene-order rearrangements or of under- or over-dosages of genes. However, it may also be that chromosomal aberrancies are the result of transformation to the malignant state, not its cause.

Development defects and aberrant chromosomes or abnormal numbers of chromosomes are also frequently associated.[19] Well-known instances include Down's syndrome and Klinefelter's syndrome, which are characterized by an extra chromosome and genic imbalances. These aberrancies are most likely the result of non-disjunction, sometimes a result of irradiation. Examination of late abortions often reveals a higher than expected incidence of chromosomal defects of the type associated with radiation exposure. Moreover, it is true that irradiation increases abortion frequency and the incidence of developmental defects. Still, unequivocal proof linking radiation-produced chromosomal aberrations and defects of embryonic and fetal development does not exist, and it can only be said that developmental defects are a possible result when radiation produces chromosomal aberrations either in germ or embryonic cells.

Finally, mitotic death is probably a major mechanism of cell killing during radiotherapy.

SUMMARY

1. Chromosomes consist of DNA (deoxyribonucleic acid) and protein.
2. Chromosome arms are genetic material arranged in a linear sequence.
3. Chromosome centromeres attach to the mitotic spindle and draw chromosomes into new daughter cells at division.
4. Radiations break chromosomal arms. Unhealed breaks can leave chromosomal fragments without spindle attachment, and at division these fragments are not included in daughter nuclei.
5. Broken chromosome arms may rearrange before healing, altering the linear order of genes.
6. Loss of genetic material from daughter nuclei at cell division can be lethal, and chromosomal rearrangements may alter genic expression.
7. Radiation injury of the centromere may cause loss of genetic material, which may be lethal, or genic imbalances, which decrease viability.
8. The frequency of chromosomal breaks and rearrangements are functions of radiation dose, LET, and dose rate and are influenced by the presence or absence of oxygen.

9. Chromosomal breaks and rearrangements are associated with both cancers and defective embryonic and fetal development.

REFERENCES

1. Bahn, E.: Position-effect variegation for an isoamylase in *Drosophila Melanogaster*. Hereditas, 67:79–82, 1971.
2. Wallace, B., and Kass, T.L.: On the structure of gene control regions. Genetics, 77:541–558, 1974.
3. King, M., and Wilson, A.C.: Evolution at two levels in humans and chimpanzees. Science, 188:107–116, 1975.
4. Dobzhansky, Th.: Genetic entities in hominid evolution. *In* Classification and Human Evolution. Edited by S.L. Washburn. Chicago, Aldine, 1963, p. 347.
5. Mayr, E.: The taxonomic evaluation of fossil hominids. *In* Classification and Human Evolution. Edited by S.L. Washburn. Chicago, Aldine, 1963, p. 332.
6. Simons, E.L.: Primate Evolution. New York, Macmillan, 1972.
7. Simpson, G.G.: Principles of Animal Taxonomy. New York, Columbia University Press, 1961.
8. Jones, K.W., et al.: Satellite DNA, constitutive heterochromatin, and human evolution. Symp. Med. Hoechst., 6:45, 1973.
9. Henderson, A., Warburton, D., and Atwood, K.C.: Localization of rDNA in the chimpanzee *(Pan troglodytes)* chromosome complement. Chromosoma, 46:435–441, 1974.
10. White, M.J.D.: Chromosomal rearrangements and speciation in animals. Annu. Rev. Genet., 3:75–98, 1969.
11. Wilson, A.C., Maxson, L.R., and Sarich, V.M.: Two types of molecular evolution. Evidence from studies of interspecific hybridization. Proc. Natl. Acad. Sci. USA, 71:2843–2847, 1974.
12. Wilson, A.C., Sarich, V.M., and Maxson, L.R.: The importance of gene rearrangement in evolution: Evidence from studies on rates of chromosomal, protein, and anatomical evolution. Proc. Natl. Acad. Sci. USA, 71:3028–3030, 1974.
13. Prager, E.M., and Wilson, A.C.: Slow evolutionary loss of the potential for interspecific hybridization in birds: A manifestation of slow regulatory evolution. Proc. Natl. Acad. Sci. USA, 72:200–204, 1975.
14. National Council on Radiation Protection and Measurement, (NCRP) Report 64: Influence of Dose and its Distribution in Time on Dose-Response Relationships for Low LET Radiations. Washington, D.C., April 1980.
15. Visfeldt, J.: Chromosome aberrations in occupationally exposed personnel in a radiotherapy department. *In* Human Radiation Cytogenetics. Edited by H.J. Evans, W.M. Court Brown, and A.S. McLean. Amsterdam, North-Holland Publishing Company, 1966, pp. 167–173.
16. Buckton, K.E., Dolphi, G.W., and McLean, A.S.: Studies of chromosome aberrations in cultures of peripheral blood from men employed at UKAEA establishment. *In* Human Radiation Cytogenetics. Edited by H.J. Evans, W.M. Court Brown, and A.S. McLean. Amsterdam, North-Holland Publishing Company, 1966, pp. 174–182.
17. Wald, N., Koizumi, A., and Pan, S.: A pilot study of the relationships between chromosome aberrations and occupational external and internal radiation exposure. *In* Human Radiation Cytogenetics. Edited by H.J. Evans, W.M. Court Brown, and A.S. McLean. Amsterdam, North-Holland Publishing Company, 1966, pp. 183–193.
18. Pitot, M.C.: Fundamentals of Oncology. New York, Marcel Dekker, 1978, pp. 60–61.
19. National Academy of Sciences. Research Needs for Estimating the Biological Hazards of Low Doses of Ionizing Radiation. Washington, D.C., 1974.

Chapter 5

Radiation Carcinogenesis

5.1 INTRODUCTION

It has been recognized for more than 70 years that exposure to ionizing radiations increases cancer incidence; that cancer incidence is elevated in both partially and totally irradiated individuals; and that in situations of partial body irradiation, tumors almost invariably appear at the irradiated site. In fact, more is probably known about the carcinogenic effects of ionizing radiations than about any other environmental carcinogen. There is a wealth of epidemiologic information gleaned from studies of exposed human populations, experimental literature, and the beginnings of an understanding of the mechanisms by which radiation damage results in cancer induction.

Most information relating radiation exposure of humans to the induction of cancer[1,2] comes from studies of survivors of the Japanese atomic bomb blasts in 1945; x irradiation used as therapy, principally for benign diseases; medically and occupationally exposed individuals irradiated with internal emitters; and radiologists exposed as part of their work. As a result of these studies there is sufficient knowledge to state that induction of malignancies probably represents the most important effect produced by exposure of the human population to low radiation doses.

Because radiation-induced cancers and spontaneously occurring ones are identical, radiation carcinogenesis is detectable only by statistical means (the presence of a cancer in any given irradiated person can never be attributed with certainty to radiation). That is, if cancer appears in greater incidence in an irradiated population than in a closely matched unirradiated one, radiation can be said to be responsible for the excess—provided all other causes are ruled out. However, among those that appear, those induced by radiation and those that occur spontaneously cannot be distinguished. Furthermore, the smaller the radiation dose to which a population is exposed, the smaller will be the likelihood that radiation-caused cancer will appear in any given irradiated person. Since low doses of radiation produce so few cancers, it is often, if not always, impossible to be sure that any at all have been induced. Most convincing information relating cancer induction in humans to irradiation comes, therefore, from exposure in the high-dose range (> 50 to 100 rads). In the dose range comprising only a few rads, the dose range to which most individuals are exposed, cancer incidence can only be *inferred* from the carcinogenic action of high doses, a procedure not without its problems. Thus, the information most urgently required, the carcinogenic potential of exposing human populations to low doses of radiation, is also the least secure.

5.2 RADIATION AS A CARCINOGEN

A carcinogen may be viewed as an agent which causes cancers that might not have otherwise developed. In laboratory animals and man, cancers develop "spontaneously" in the apparent absence of exposure to any known carcinogen. All types of cancer that result from exposure to carcinogens also occur spontaneously, even if the spontaneous incidence is low. A carcinogen may, therefore, be defined as an agent that *increases the risk* of development of one or more forms of cancer compared to the risk of development of the same cancer(s) in the absence of exposure to the agent.

Experimental exposure of animals to radiations and observations of exposed human populations have shown that ionizing radiations are *general* carcinogens, capable of inducing tumors in almost all tissues of mammals irrespective of species. None of the chemical carcinogens is so universal in its actions.

5.3 LATENT PERIOD

Considerable time may elapse between radiation exposure and the appearance of cancer. This period has been called the "latent period." Some feel that no cancer is present then and that irradiation produces only a precancerous state in some cells which must have some other stimulus to begin malignant growth. Others maintain that radiation transforms very few cells to the malignant state, and since there are so few at the start, years must pass before enough accumulate to be detectable and recognized as cancers.

Where radiation is the carcinogen, the length of the latent period is usually long (relative to the life span of irradiated individuals). Solid cancers generally have longer latent periods than hematogenous cancers; they seldom appear before 10 years

after radiation exposure, have *mean* latent periods of about 20 years, and may continue to appear for 30, 40, or more years following exposure. Leukemias, on the other hand, generally appear within 2 to 4 years after radiation exposure and usually disappear within 30 years (Table 5.1). The projected period within which the carcinogenic and/or leukemogenic action of in utero radiation exposure are thought to be expressed is about 10 years. These long latent periods complicate both prospective follow-up of irradiated populations and retrospective evaluation of cancer patients for a history of irradiation. If *mean* latency is about 25 years, then the total cancers diagnosed within this time would only represent about half of all cancers likely to be induced in young subjects with a long life expectancy.[2]

5.4 MECHANISMS OF RADIATION CARCINOGENESIS

The precise mechanism by which radiation transforms normal cells to cancerous ones is not known at this time; however, most experts studying the problem seem to agree that carcinogenesis is the result of *somatic mutations*, genetic changes in the DNA of normal cells or changes in the *expression* of the genes of normal cells. The idea that chromosomes have something to do with cancer is almost as old as the dis-

Table 5.1. *Approximate latent periods for radiation-induced cancers in human beings in years*[2,3]

Type	Minimum	Mean	Total Period of Expression
Leukemia	2–4	10	25–30
Bone	2–4	15	25–30
Thyroid	5–10	20	>40
Breast	5–15*	23	>40
Other Solid Tumors	10	20–30	>40

*Varies with age at exposure.

covery of chromosomes. In the late 1890s, altered chromosome patterns and mitotic abnormalities, as well as irregularities in cell division, were observed in tumor cells. Theodore Boveri, at the beginning of this century, formulated his "chromosomal theory of malignant tumors," which, in summary says: Tumor cells have chromosome numbers which deviate from the normal. This was soon followed by the mutation theory of malignant growth: Carcinogenic stimuli (such as irradiation) cause mutations; mutations in the growth-regulating portions of the genetic material are responsible for the neoplastic properties of tumor cells.

If the mutation theory is correct, good mutagens should be good carcinogens, and vice versa. It has long been known that x rays and many other agents or substances *are* both carcinogenic and mutagenic. There are other factors supporting the mutation hypothesis:

1. Cancers are composed of malignant cells of given tissues. Liver cancers are made of cancerous liver cells; breast cancers of breast cells. Even when metastasized to other organs, they retain characteristics of the tissue of origin. For example, breast cancer spread to bone remains recognizable as having originated in the breast.

2. Cancer cells "breed true," i.e., they give rise to other cancer cells like themselves. This suggests that they have a stable genotype.

3. In a large majority of cases, the conversion of a normal cell to a cancer cell is irreversible. It is a road of no return. This is consistent with what is expected from mutation.

4. Changes from normal chromosome patterns are often seen in tumors. Occasionally certain neoplasms are distinguished by a characteristically altered chromosome. A well-known example is the Philadelphia chromosome in chronic myeloid leukemia.

5. It is known that in certain rare human diseases (Fanconi's syndrome, Bloom's syndrome and others) there is a predisposition both for increased chromosome aberration frequency and for a greatly increased incidence of leukemia.

6. Altered genetic material which leads to tumors as a final consequence can be isolated by breeding. Animal strains thus obtained can have a 100% incidence of certain spontaneous tumors.

Viruses are known to be associated with cancer cells, especially in lower animals, and with an increased incidence of cancer. It has been suggested that radiations' carcinogenic action may be to stimulate or release a dormant oncogenic virus. The incorporation of the viral genome into a cell's genome would constitute a change in the DNA of the cell; that is, a mutation.

There is disagreement over whether cancers arise from a single original transformed cell or whether several such cells are needed to begin cancers. Two observations point to the possibility of single-cell origin. First, in women who are heterozygous for electrophoretic variants of X-linked, glucose-6-phosphate dehydrogenase, cancers are uniformly of one phenotype or the other, whereas a comparable amount of normal tissue is composed of a mixture of the two phenotypic classes.[4] Second, there is evidence for the single-cell origin of cancers from experiments in which transformed cells are transplanted into animals. Although there is a controversy associated with various aspects of this approach, it does seem to show that a single cell may give rise to a cancer. Others[5] have presented evidence, however, that as a result of *low* doses of radiation, tumor formation may be the end point of damage to a collection of irradiated cells. The researchers feel that damage to a single cell is not sufficient to account for *radiation* induction of malignancies.

Even though there is evidence for single-cell origin of cancers, a single classical mutation is probably not enough to produce cancers. Many favor a "multi-event" concept in which the induction of latent tumor cells is believed to be caused by a carcinogen and has something to do with damage to the genetic material (initiation). The latent tumor cells that result are "developed" into a cancer by agents called "promotors." Promotors may or may not be carcinogens. Examples of promotors would be thyroid stimulating hormone and mammotrophic hormone, neither of which is a carcinogen, in the development of thyroid and breast cancer. Some promotors have been shown to be stimulators of DNA synthesis, and this, in fact, may be the mode or one of the modes of action of some or all of these agents.

There is also strong evidence that the body's immune system carries on a continuous surveillance for transformed cancer cells, which are recognized as foreign and are destroyed. Radiation in appreciable doses can suppress immune responses. Conceivably, immune surveillance is depressed by irradiation, and this allows the growth of cancer cells that otherwise would have been killed. Perhaps the long latent period is related to development of foci of quiescent cancer cells held in check for many years by immune suppression. Subsequent changes in the immuno-competence of the host through illness, drugs, or other means, may free these to grow. Evidence supporting the immune surveillance idea includes the observation that cancer incidence is higher in people with less effective immune systems than those with strong response. Growth of latent tumor cells may also be related to a process imaginatively called "sneaking through." Malignant cells may be found or located in a tissue compartment that is for some reason sheltered from immunologic surveillance or attack. They may multiply, and tumor cell number may pass the threshold

of reversibility before enough antigen is released to mobilize the immune system.

While the mechanism of radiation carcinogenesis is fairly well accepted as a disturbance in the genetic apparatus, other ideas carrying substantial weight are tenable.

—

5.5 DETECTION OF RADIATION-INDUCED CANCERS

The presence of radiation-induced cancers in a human population is difficult to detect and measure for the following reasons:

1. Cancers induced by radiation are indistinguishable from those occurring naturally; their existence can be *inferred* only on the basis of a statistical excess above the natural incidence.
2. The incidence of spontaneous or naturally occurring cancers is quite high. This high background makes it difficult to detect small excesses. For example, the naturally occurring *lifetime* risk of *fatal* cancers[1] is about 160,000 cases per million persons, a rate of 16%. Thus, very large populations are required to provide convincing evidence of radiation-induced cancers, particularly if the radiation doses are low.
3. The long latent period complicates follow-up of irradiated populations as well as retrospective evaluation of cancer patients for a history of irradiation.
4. In several instances data have been derived from therapeutically irradiated patients in whom the radiation effects may have been complicated by the effects of the disease itself or by the effects of other treatments.
5. The excess incidence of malignant disease in an irradiated population can be reliably estimated only by comparison with a control population

which is similar in all respects, except that it has not been irradiated. This condition is seldom fully achieved.

5.6 DOSE AND CARCINOGENESIS

Contrary to a certain popular opinion, exposure to high doses of radiation does *not* imply almost certain cancer induction. In fact, high doses of radiation (>100 rads) to the whole body are known to cause only a *small* increase above the background cancer rate. For example, comparisons of 82,000 Japanese atomic bomb survivors against age- and sex-matched controls have shown (through 1974) 3,842 cancer deaths among 19,646 total deaths in exposed survivors. In this heavily irradiated population, dose-response analysis suggests that only about 85 leukemias and 100 solid tumors were attributable to bomb radiation.[6] Thus, the likelihood that radiation cancer will be induced is not great, even if doses are high; in a population receiving 100 rads of whole-body radiation, the incidence of *induced* fatal cancers over the lifetime is about 1%.

5.7 DETECTION OF INDUCED CANCERS AT LOW DOSES

It is difficult to estimate directly the risk of cancer induction by radiation doses of 10 rads or less, because the degree of risk is so low, few cancers are induced against a background of large numbers of spontaneous cases. The difficulty in recognizing excess cancers induced by radiation can be appreciated from the following example: 1000 women who received 100 rads to their breasts developed 40 cancers where 22 spontaneous cancers were expected.[7] This difference, 18, is highly significant and can be used to estimate the "expected" effects at low doses. A linear or proportional approach would imply that, if the dose were reduced by a factor of 100 (to 1 rad) and

the population increased by the same factor (to 100,000), the same number of radiation-induced cancers (an excess of 18) should be induced. However, the expected number of spontaneous cancers (the number in a matched control population of the same size) in the larger population is no longer 22, but 2200. The difference (2200 vs 2218) is questionable and very difficult, if not impossible, to validate statistically. Clearly, to obtain direct information about carcinogenic effects of low doses of sparsely ionizing radiation, sample sizes that are impractically large are required. The smaller the radiation dose, the smaller the anticipated excess cancer risk. As the excess risk decreases, progressively larger samples are required in order to detect it. For example, if excess risk is proportional to radiation dose, and if a sample of 1000 exposed and 1000 control subjects is necessary to determine the carcinogenic effect of a 100-rad exposure, a sample of 100,000 each may be needed to determine the carcinogenic effects of a 10-rad exposure and about 10 million in each group to analyze effects of 1 rad.[8] Thus, conceivably, it may not be possible ever to observe directly the carcinogenic effect of radiation doses in the medical, diagnostic range. Given these problems, the following section describes how, in the absence of direct observations, estimates of the carcinogenic effects of low radiation doses are in fact made.

5.8 THE DOSE-EFFECT RELATIONSHIP

Prior to about 1950 it was assumed that for cancer induction as for other somatic effects, such as cataracts and sterility, there was a threshold dose below which no cancers were induced. With the accumulation of more data on the induction of genetic effects and on the association between carcinogenesis and mutagenesis, radiation carcinogenesis began to be seen as an effect for which there is no threshold of induction. As stated earlier, risk estimates for a variety of human cancers are fairly well established at high doses (>100 rads) but there is little *direct* evidence of the degree of risk at lower doses, particularly under 20 rads. In order to estimate what risk there may be from low-dose radiation exposure, it becomes necessary to "extrapolate," actually to interpolate, between cancer incidences following exposure to high doses of radiation and the spontaneous rate that occurs in the absence of any except background radiation exposure. The process of interpolation requires the *assumption* of a specific relationship between dose and effect so that a curve or line can be drawn between cancer incidence occurring after high doses and the spontaneous rate. The assumptions made and how this curve is drawn (the shape of the curve) are at the center of the controversy that surrounds estimation of low-dose effects.

The general approaches to the interpolation process are shown in Fig. 5.1. Appropriate data which exist at high doses are used to interpolate back to the zero-dose incidence in three principal ways: linear, above linear, and below linear.[9] Before considering each of these models in detail it should be pointed out that both human data and animal experiments relating tumor incidence to dose eventually reach a dose range in which tumor incidence no longer rises but levels off (the efficiency of carcinogenesis drops), and tumor incidence may actually drop after a peak is passed—the peak occurring generally at about a 30 to 40% tumor incidence (Fig. 5.2). Thus, the models (linear, above linear, below linear) may need to be modified to allow for the tendency of higher doses to kill cells rather than make them cancerous.

The Linear Model

Most of the present radiation safety standards for occupational exposure *assume* that the dose-response curve is *linear* (see Fig. 5.1), possesses no threshold, and that reduction of dose or dose rate does not reduce risk. This direct or straight-line relationship predicts that, *per rad*, low doses would have the same probability of inducing cancer as high doses. According to the linear hypothesis, if a dose of 100 rads given to 1000 people produces 20 extra cancers, then a dose of 1 rad to 100,000 people would produce the same number of cancers (20). For high-LET radiations (such as neutrons), a linear dose-response appears to fit data from animal experiments and human exposure.[10] However, most experience in induction of mutation and carcinogenesis by low-LET radiation indicates that the response curve is not linear but curvilinear as in Fig. 5.2. If the actual dose-response curve for radiation-induced cancer in humans is, in fact, curvilinear, then expectations of damage based upon a linear, no-threshold model would overestimate the *risks* of low doses and low dose rates from low-LET radiations. For this reason, estimates of risk using the linear, no-threshold model have usually been regarded as upper limits of risk.

The Below Linear Model

According to this model, the effectiveness of a given quantity of dose for cancer induction is less in low-dose ranges than in high-dose ranges. The basis of this curvilinear model is that energy deposition is required in *two* distinct intracellular sites for an effect to be observed. Hence the probability of an effect depends upon a dose-squared (D^2) or quadratic term.

The Quadratic Below Linear Model

The pure quadratic (dose-squared) model is characterized by a slope of zero at zero dose and a *continuously* increasing slope, i.e., *risk per rad* increases continuously all along the dose axis.

The Linear-Quadratic Below Linear Model

The most widely accepted *below linear* model is the so-called linear-quadratic (LQ) model.[12,13] The LQ model for low-LET radiation actually combines aspects of both the linear and quadratic forms. There is

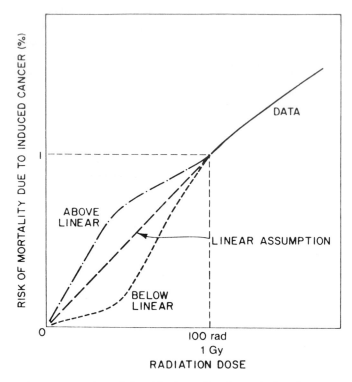

Fig. 5.1. Diagrammatic representation of the *risk of mortality* due to induced cancer versus radiation dose from low-LET radiation. Data at higher doses (generally above 100 rads) allow the estimation of risk with some precision. From Sinclair, W.K.[9]

an initial linear portion with no threshold followed at higher doses by a quadratic term (Fig. 5.3). This sort of response, the LQ model, can be represented by the expression:

$$I = aD + bD^2 \qquad (1)$$

where I is the incidence, D the dose, and a and b are constants. The quadratic term (bD^2) predominates at higher doses (perhaps above 50 to 100 rads) and at high dose rates, whereas the linear term (aD) and the slope it represents are dominant as dose and dose rates are reduced. A relationship such as this is based upon the observation that the biologic effect from a given dose of low-LET radiation is usually dependent on the time rate of energy deposition or on the total elapsed time during exposure (reduced dose rates or fractionation usually produce a reduced effect).

The curvilinear relationship and dose-rate effect observed using low-LET radia-

tions, in contrast to the fully linear response and lack of dose-rate effect observed using high-LET radiations, led to the concept of "dual radiation action" formulated by Kellerer and Rossi.[14,15] This model, a generalized two-hit model for permanent damage which is based upon microdistribution of energy deposited in cells, postulates that most low-LET radiation interactions in small intracellular sensitive sites or volumes produce reparable sublesions. These are transformed into permanent biologic lesions (such as carcinogenic transformations) only if second hits or sublesions occur *close* to the first ones in both time and distance. Since each sublesion is usually produced by independent ionizing particles, the number of *permanent* lesions (two hits close in time and distance) depends on the square of the dose (bD^2). However, enough energy can be deposited within intracellular sensitive sites by indi-

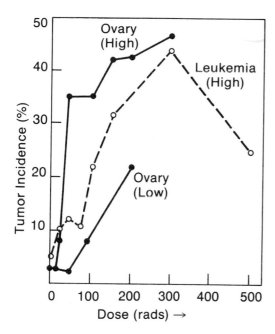

Fig. 5.2. Incidence of tumors induced in mice by external low-LET radiations. The dose rates are indicated on the figure as "high" or "low" for ovary tumor[10] and leukemia[11] induction. Note the reduced tumor induction efficiency at about 30 to 40% incidence. The response for ovarian tumors is curvilinear with a marked reduction in effectiveness at lower dose rates.

Fig. 5.3. The data at high doses (data points are represented as closed circles) can be assumed to fit a linear (A) or a linear-quadratic (B) relationship. The linear-quadratic relationship is represented by an initial linear portion followed at higher doses by a quadratic portion. Note that if the actual relationship is linear-quadratic, the risk at low doses is overestimated by the linear form. Because of repair, the linear-quadratic is reduced to the linear component at low doses and low dose rates (C).

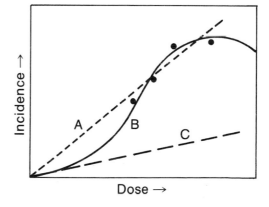

vidual high-LET radiations to produce *both* interacting sublesions at once. The number of permanent lesions resulting from high-LET radiations, therefore, is proportional to the dose (aD).

The presence of a linear component in the dose-response relationship for low-LET radiations implies that they do some damage equivalent to high-LET radiations (both sublesions produced by one particle and little or no repair occurs). As a matter of fact, all low-LET radiations, such as the electrons produced by x rays or gamma rays, have as part of their track, high-LET components, with the high ion density occurring at the ends of their tracks. Thus, the linear-quadratic model for low-LET radiations postulates two different portions: a linear dose response portion, occurring primarily in the low-dose range, followed by a predominantly quadratic portion occurring at higher doses, each having its own mechanism of damage production. The initial linear portion represents irreparable damage produced by intense ionization from high-LET portions of low-LET electron tracks. This damage would be unaffected by dose rate. The quadratic portion represents the formation of sublesions by more diffuse ionizations. If such sublesions are capable of repair, the number of permanent biologic lesions formed would depend upon the number of sublesions present in a given volume at a given time. This means the number of permanent lesions would depend upon *both* the total dose and the dose rate. When both dose and dose rate are high, large numbers of lesions would be produced per unit volume of matter, and there would be insufficient time to allow repair to reduce their numbers. If dose were reduced, even if dose rate remained high, fewer lasting lesions would be produced since fewer sublesions would be available for interaction.

At low doses, permanent lesions would be formed almost entirely by the passage of densely ionizing tails of low-LET radiations' tracks through sensitive sites, and

dose response would be linear, with effect proportional to dose (see Fig. 5.3). If, on the other hand, dose rate were progressively lowered while doses were held constant, even at higher doses, accumulation of sublesions would be expected to be slower since there would be greater opportunity for repair between the formation of each one (fewer sublesions formed per unit time). If dose rate were low enough, there would be time for repair of every sublesion, and permanent lesions would again be formed almost entirely by single events (i.e., the passage of densely ionizing tails of low-LET radiations through sensitive sites). A linear (proportional) relationship would again result. Extension of the initial slope of the linear portion of linear-quadratic relationships to higher doses would give the limiting effect for progressively lowered dose rates (see Fig. 5.3). A dose rate would be encountered beneath which no further reduction in production of permanent lesions would occur.

As stated earlier, it is often found in experimental carcinogenesis and mutation induction that there is a peak (see Fig. 5.2) in dose response. If dose is increased beyond the peak, mutation or cancer induction becomes *lower* per unit of dose (the simple form of Equation 1 does not contain a factor to predict this). The turning down of the curve at high doses has been attributed to the killing of cells that otherwise would give rise to mutations or cancers. This is *an* explanation for the apparent lack of induced thyroid cancers subsequent to ^{131}I treatment for hyperthyroidism. The high therapeutic radiation doses probably destroy large numbers of thyroid cells, leaving too few proliferative cells to have a high probability of producing a cancer.

Above Linear Model

A discussion of radiation carcinogenesis in humans would not be complete without mention of recent suggestions of relatively high carcinogenicity in adults exposed to low doses of ionizing radiations. Examples of such reports[16] include: the Mancuso Report[17] of excess cancer mortality among occupationally exposed workers at the Hanford Atomic Plant, in Richland, Washington, and commentary on this report;[18] the report of Bross and the tri-state leukemia survey[19] claiming supersensitivity of certain groups of subjects to low doses of diagnostic x rays and commentary on that report;[20] the report of excess malignancies among Portsmouth Naval Shipyard workers exposed to low occupational doses;[1,21] and the "Project Smoky" study indicating excess leukemia among soldiers participating in the 1957 nuclear test explosion.[22] Each of these studies suggests the *possibility* of a rapidly rising dose-response pattern at low doses (see Fig. 5.1), pointing to and perhaps related to a greater than expected sensitivity at low doses, the presence of particularly susceptible individuals in the human population or to a "leveling off" of the dose-response curve at very low dose levels due to cell killing. However, significant degrees of cell killing are not expected at such low doses. Most human and animal experience indicates that saturation, followed by significantly reduced efficiency for radiation carcinogenesis, occurs when tumor incidence reaches 30 to 40%.[9] Since, in humans, a dose of 100 rads corresponds to induced tumor incidence of only 1 to 2%, it seems unlikely that there is enough cell killing even at doses between 100 to 200 rads to cause leveling off in the dose-response relationship, let alone at the very low doses to which persons in the above cited studies were exposed.

Whether exceptionally sensitive individuals exist or not cannot be shown at this time. However, if they do, their increased low-dose sensitivity would mean that as much as 50 to 70% of the *total* of human cancers would be induced by natural background radiation; the current estimate is about 1%.[23] In fact, a recent study of Frigerio and Stowe[24] examined the relationship of background radiation levels in the United States to incidence of malignancy.

When states were compared as groups (the 7 states with natural backgrounds above 165 mrem/year, the 14 states with backgrounds above 140 mrem/year, and the 14 states with the lowest backgrounds, averaging 118 mrem/year) there appeared to be a *reverse relationship,* that is, rates of malignancy tended to decrease with increasing background. This and many similar studies done in other countries have not detected increased malignancies in areas with high background radiation and indicate that the carcinogenic effects of low doses, if present, must be small.

Most studies reporting high carcinogenesis at low doses have been questioned or refuted.[16,18,20] As the 1980 BEIR Report[1] states: "Studies by a number of scientists who have claimed a greater carcinogenic effect due to exposure to low-dose ionizing radiation than generally accepted are reviewed. . . . None of these studies was considered by the Committee to constitute reliable evidence at present for use in risk estimation, for various reasons, including inadequate sample size in some instances, inadequate statistical analysis, and unconfirmed results."

Choice Between Models

Human data on radiation-induced cancer seldom provide dose-response relationships in the low-dose range, and when they do, the reliability is such that discrimination among the possible models is usually not possible. There is also the possibility that all types of cancer induction may *not* follow the same dose-effect relationship. For example, the dose-incidence relationship for leukemia induction by gamma rays (low-LET radiations) in A-bomb survivors appears curvilinear while for x-ray-induced (low-LET radiations) thyroid and breast cancer, it is more consistent with a linear relationship.

Most of the radiation protection standards for occupational exposure to radiation assume a linear, nonthreshold dose-effect relationship, because it is felt that this model is conservative; viz., that it is either accurate or overstates the risk of low doses. The linear model was used in the 1977 UNSCEAR Report[2] to estimate carcinogenic risk of low doses. In the 1980 report the BEIR III Committee[1] chose, however, to adopt the linear-quadratic dose-effect relationship for low-LET radiation as the "preferred model" for estimating the total cancer risks from whole-body irradiation, since they felt this model "to be consistent with epidemiological and radiobiologic data, in preference to more extreme dose-response models such as the linear and the pure quadratic." The Committee decided that since the linear-quadratic relationship is intermediate between the linear and pure quadratic models, the latter two could be used to define the upper and lower limits of their estimates, respectively.

5.9 HOW IS RISK EXPRESSED?

Estimates of risk from radiation exposure can be expressed in *absolute* terms (which estimates the number of cancers induced per unit of radiation dose in an exposed population) or in *relative* terms (comparison, usually as a ratio, of the risk of cancer in an irradiated population to the risk of cancer in a similar, but unirradiated population).

The *absolute risk* is the excess risk due to irradiation and is, *in practice*, usually expressed as the difference between irradiated and unirradiated populations. The absolute risk may be expressed as the number of *excess* cases of cancer occurring, per unit of time, in a population exposed to a given unit of dose (for example, 1 case per million irradiated people per year per rad), or in terms of lifetime expectancy (say, 25 cases expected among a million people irradiated with one rad).

The *relative risk* for subjects exposed to a given mean dose is the ratio between the risk of cancer induction in irradiated and unexposed populations and is usually ex-

pressed as a fraction or a multiple of the spontaneous or natural incidence or risk. The doubling dose, the dose that doubles the natural incidence, is an example of a special calculation of relative risk. Relative risk values can be somewhat misleading if the natural incidence of a particular cancer is quite low. Even if a dose of radiation doubles the risk in a population it may, in fact, indicate *few* new cases.

5.10 ORGAN SENSITIVITY TO RADIATION CARCINOGENESIS

Different tissues of the human body apparently respond differently to radiation. Some tissues are somewhat more sensitive than others to the induction of cancer. Recent estimates of the risk of *fatal* radiation-induced malignancies are given in Table 5.2. Although they were derived by different scientific study groups, the results are very similar. It should be noted that risk of radiation-induced cancer can be expressed either as cancer *mortality* or cancer incidence. The major differences between estimates of radiogenic cancer incidence and mortality depend on two sites, thyroid and breast. Thyroid is sensitive to radiation car-

Table 5.2. *Estimates of lifetime risk of fatal radiogenic cancer for low-LET radiations*[2,25]

Types of Cancer	Cases per million persons per rad per site		
	UNSCEAR, 1977		ICRP 26
Female Breast	50	(100)	25
Thyroid	10	(100)	5
Lung	25–50		20
Leukemia*	20–50		20
Stomach, liver, colon	10–15		5
Bone, esophagus, small intestine, bladder, pancreas, rectum, lymphatics	2–5		5

( ) Incidence, rather than fatal cases.
*All forms of leukemia except chronic lymphocytic.

cinogenesis but has a low associated mortality, because the type of thyroid cancer induced by radiation (adenocarcinomas arising from follicular cells) has an unusually slow progress and a high cure rate. The female breast has a similar radiogenic incidence but a moderately high mortality with about a 50% cure rate. For the remainder of the radiation-induced cancers, the difference between incidence and mortality is not substantial. Table 5.2, which gives the *absolute risk* (for cancer incidence), shows that breast and thyroid are well ahead of bone marrow and lung. However, based upon *relative* risk (the fractional increase over the spontaneous level), the sequence is probably thyroid, bone marrow, lung, and breast. Because of the high relative risk of leukemia and its short latent period (see Table 5.1), it was initially thought to be the dominant form of radiation-induced cancer. With longer observation times after exposure, in groups such as the A-bomb survivors, the leukemia risk has fallen off, while mortality from solid tumors has continued to increase.[6] It now appears that a ratio of about 4:1 between the frequency of other malignancies and leukemia is likely.[30]

Effect of Gender

The incidence of radiation-induced human breast and thyroid cancer is such that the *total* cancer risk is higher for women than for men. Breast cancer occurs almost exclusively in women and is very sensitive to radiation induction. Thyroid cancer induction by radiation is about three times higher for women than for men (the natural incidence is also higher). There is some evidence that males may be slightly more susceptible to radiation-induced leukemia than females. With regard to other cancers, there is little difference in the radiation risk related to gender.

Effect of Cell Division and Hormones

Induction of cancer by radiation does not depend upon rate of cell division. If this were true, the small intestine should be as sensitive as the bone marrow, which is not the case. Sensitivity is also not *necessarily* related to the influence of hormones. Pituitary or sex hormones are important in the natural occurrence of cancers of the female breast, thyroid, uterus and prostate, yet female breast and thyroid are quite sensitive to radiation-induced cancer while uterus and prostate are not.[1]

Effect of Age

Since cancer is primarily a disease of old age, age is the most important factor influencing risk of *spontaneous* cancer. The influence of age is not uniform in human radiation carcinogenesis. The clearest evidence of a very high risk for those irradiated in early life is found in the leukemia experience of A-bomb survivors (the induction rate was highest for those under 10 and those over 50). Women exposed in the second decade of life (when major hormonal changes and breast development occur) appear to be at the highest risk of breast cancer. Tumors of lung, stomach, and intestine have shown increased risk with increased age at exposure, but there is *no* strong evidence of an age relationship for thyroid cancer.[1] There is also evidence for greater radiosensitivity during prenatal life judging from incidence of childhood cancer and leukemia following in utero irradiation.[26]

Effect of Other Factors

Some groups in the general population appear to be at increased risk of induction of cancer by radiation. Examples of these groups include Jewish children when compared to other ethnic groups for induced thyroid cancer;[27,28] and excess lung cancer among uranium miners who were cigarette smokers.[29]

5.11 TOTAL LIFETIME CANCER RISK FROM RADIATION

Variables such as tissue and age sensitivity, latency, duration of effect and questions about dose level and fractionation make it difficult to quantify the *total* cancer risk. Lifetime absolute risks following whole-body exposure at low doses (less than 10 rads) cannot be derived with confidence from risks at high doses. Estimates of effects following low doses are dependent upon choice of a dose-response function (e.g., linear vs linear-quadratic). Human data simply are not adequate to permit discrimination between the possible choices. The reports issued by UNSCEAR[2] and the ICRP[24] use a linear, no-threshold relationship. The BEIR III report[1] uses the linear-quadratic model as its "preferred" model, although estimates of risk for low-LET radiation were also given for the linear and pure quadratic models, providing upper and lower limits respectively. Comparative estimates for lifetime risk of fatal cancer induced per rad for a population of 1 million are given in Table 5.3. It is apparent that, even with differing dose-effect relationships, the agreement is good, centering around a value of 100 lifetime fatal malignancies (with an upper limit of about 200) induced by a dose of 1 rad in an exposed population of 1 million, as compared to spontaneous lifetime cancer risk of about 160,000 cases per million persons. Thus in a population of 10,000, one would normally expect 1600 cases of fatal cancer; exposure of each of the 10,000 persons to 1 rad of low-LET, whole-body radiation would be expected to increase those deaths to 1601.

Table 5.3. *A comparison of additional cancer deaths in the lifetime of 1 million people per rad after exposure to a few rads of low-LET radiation*

Number	Source
77–226*	BEIR, 1980
75–175†,‡	UNSCEAR, 1977
100†	ICRP 26, 1977

*Linear-quadratic model.
†Linear model.
‡Induced nonfatal malignancies estimated to be about the same magnitude.

Most of the human experience upon which such estimates of cancer risk are based involves four major irradiated groups:

1. Survivors of atomic bomb blasts at Hiroshima and Nagasaki in 1945
2. Patients exposed to x irradiation either for therapy for benign conditions or for diagnostic radiologic studies (particularly antenatal x rays)
3. Patients who were irradiated with internal emitters
4. Persons exposed occupationally to radiation

Table 5.4 summarizes the types of cancer linked to radiation by studies of these groups. The association is stronger in some than in others. Strong associations usually are confirmed by multiple studies with evidence that risk increases with dose. Weaker associations usually are single sets of data which are unconfirmed by other studies. As mentioned earlier, such epidemiologic studies have inherent problems which include long latent periods (since these studies are retrospective, exposure to radiation occurred in the distant past, and exact dose to particular individuals or groups may not be known), and the complicating factor of disease (generally the group has been exposed for a medical reason and may not be representative of the population at large). Given these shortcomings we will briefly describe the circumstances of irradiation and the findings in these four groups.

5.12   ATOMIC-BOMB RADIATION

Japanese A-bomb Survivors

The Japanese A-bomb survivors are the only human population in which relatively large numbers of normal individuals of all ages were exposed in the same manner to a range of radiation doses, all at the same time. For this reason they are regarded as the most important source of information on the late somatic effects of ionizing ra-

Table 5.4. *Cancers associated with irradiation of particular populations*[23]

Type of Cancer	Atom Bomb Radiation		Medical X Ray						Medical Radionuclides				Occupational Radiation		
	A-bomb survivors[1,2,6,8,30]	Marshall Islanders[1,2]	Ankylosing spondylitis[33]	Mastitis[34]	Chest fluoroscopy[34]	Castration[35]	Thyroid[32,36-38,40]	In utero[26,42-45]	Thoratrast[54]	Radium injections[52,53]	131I for thyroid ca[58,59]	32P therapy[58]	Radiologists[1,60-64]	Radium dial painters[52]	Uranium miners[28]
Leukemia	●●●		●●●			●●	●●	●●●	●●●		●	●	●●●		
Thyroid	●●●	●●					●●●								
Breast	●●●			●●●	●●●							●			
Lung	●●●		●●●												●●●
Bone										●●●				●●●	
Liver									●●●						
Skin						●●							●●●		
Lymphoma	●●												●●		
Esophagus	●●		●●												
Stomach	●●		●●												
Bladder	●●														
Colon						●●								●	

●●● Strong association, confirmed in multiple studies, with evidence that risk increases with dose.
●● Meaningful association, but less striking.
● Suggestive association, usually unconfirmed.

diation in humans. Several features make this group extremely useful[1]—it is the largest of its kind; its members were not selected according to disease or working status; intensive follow-up has been under way for 30 years, enhanced by the Japanese family registration system which virtually guarantees 100% mortality follow-up; and an elaborate dosimetry system exists which estimates total body and organ doses. There are also limitations and problems—although the samples are large, they are not of sufficient size to detect with high probability the effects at low doses; the exposure rate was very high, and the radiation was a mixture of gamma rays and neutrons; effects of heat and blast complicate the study of effects; living patterns were disrupted in the devastated cities; and there is a possibility that the survivors may have been "fitter" than those who died from the bomb blast.

At the time of the 1950 national census in Japan, 284,000 residents were identified, and from this group 109,000 subjects were chosen for the extended Life Span Study (LSS) of whom 82,000 survived the blasts, while another 27,000 were residents but not in the cities (NIC controls) at the time of the blasts.[6,8] Structural differences in the bombs dropped on Hiroshima and Nagasaki were thought to have resulted in different types of radiation exposure in the two cities. The major radiations were gamma rays, but the Hiroshima bomb was thought to have a very significant neutron component. Recent studies indicate that the radiation estimates for the Hiroshima bomb were in error, that the neutron component was nearly negligible, and that the dose from the gamma rays was higher than the original estimate. Preliminary calculations indicate that although the risk estimates for higher gamma-ray doses may increase, the risk coefficients for low doses would be decreased when compared to the previous estimates.[31] Thus, at low doses where population risk centers, the risk estimates, if changed, would likely be reduced by a small amount.

The most important later effects in the A-bomb survivors have been the occurrence of certain types of cancer, particularly leukemia and cancers of breast, lung, and thyroid. Weaker associations between radiation exposure and increased occurrence of lymphoma, multiple myeloma, and cancers of the stomach, esophagus, urinary bladder, and salivary glands are now under intensive investigation. The increased incidence of radiation-induced leukemia, which began to be apparent about the second year after exposure, is essentially over. On the other hand, the increase in incidence of radiation-induced solid tumors, which commenced in the late 1950s and early 1960s, has continued its upward trend and is expected to continue as the A-bomb survivors reach the age range at which cancer susceptibility is greatest.[30]

One final note—even in this study, the identification of appropriate control subjects has been a problem. Use of mortality rates for the Japanese general population presents problems[2] because living conditions in the two cities, obviously difficult for long periods of time after the explosions, could influence mortality statistics; the populations exposed were particularly depleted of healthy men of military age; and the incidence of some cancers (bronchus) differs between urban and rural populations. Use of the not-in-the-city (NIC) population as a control has also been criticized since these individuals included more immigrants, with a different medical history, than permanent residents. This has led to proposals that the so-called "zero"-dose population, actually those present during the blasts but who received only 0 to 9 rads, may be a more "valid" control population. Yet, there is a major fault in using both NIC groups and 0- to- 9-rad groups as controls—there were relatively small numbers in these groups, making both mortality and cancer incidence statistics imprecise.

Marshall Islands Fallout

In March 1954, inhabitants of several small South Pacific islands were accidentally exposed to fallout from the test of a thermonuclear bomb.[1,32] They were exposed to external radiation and to ingested radioiodine. The approximate dose to thyroid gland from radioiodine in adults was estimated at from 220 to 450 rads. Forty of the 243 subjects developed thyroid nodules and seven developed thyroid cancer.[1,2]

5.13 MEDICAL X IRRADIATION

Ankylosing Spondylitis

Over 14,000 patients (mainly males) were treated with x rays for spinal arthritis over a 20-year period starting in 1935.[33] The treatment usually consisted of about ten treatments giving a total dose ranging from 300 to 1500 rads. Leukemia appeared within a few years after exposure, and other cancers occurred later at heavily irradiated sites (for example, lung, esophagus, stomach, and bone).

Tuberculosis and Mastitis Patients

The tuberculosis group involves women who received multiple chest fluoroscopies between 1930 and 1954 to monitor artificial pneumothorax treatment for their disease. Increased incidence of breast cancer resulted. The average breast dose for the 1047 women treated was about 150 rads.[34] Fractionation did not seem to diminish the risk of breast cancer (some received more than 75 examinations).

The mastitis group involves women with an increased incidence of breast cancer after radiation therapy for acute postpartum mastitis. The average radiation dose was 247 rads,[34] and the study used both women with nonirradiated breasts and sister controls.

The dose-response curves for these

groups (and increased incidence of breast cancer among A-bomb survivors) are compatible with a linear relationship, although other relationships cannot be ruled out. A decrease in cancer incidence following large single-breast exposure (about 400 rads) is suggested for the mastitis patients. However, it appears that a similar large dose in the tuberculosis-fluoroscopy patients (small fractions over a period of years) does not show diminished cancer response. Age at irradiation had a major influence in that women irradiated as adolescents had a higher risk of radiation-induced cancer than those irradiated when older.[34]

Cervical Cancer and Castration

Women exposed to pelvic irradiation from radium implants have demonstrated no excess leukemia or solid tumors (the time interval, however, may not yet be long enough for solid tumors to have been detected). However, a study of patients exposed to ovarian irradiation (500 to 1000 rads) for castration has indicated an elevated risk of leukemia as well as cancer of the intestines, rectum and uterus.[35] Absence of leukemia in patients treated for cervical cancer and its presence in castration patients suggests that higher doses in the former group resulted in cell destruction lowering the probability of cancer induction.

X Irradiation of the Thyroid

Beginning in the 1920s radiation therapy was used in treatment of infants and children for enlarged thymus, tonsils, or adenoids and for acute infections of the head, neck, and chest. Ringworm of the scalp, acne, and various dermatoses were also treated. In the era prior to drugs such as sulfonamides and antibiotics, these methods were found effective in the cure and control of these diseases, and prior to 1950 thyroid gland was regarded as relatively resistant to radiation-induced cancer.

As a result of these practices, thyroid gland frequently received either direct or scattered radiation (the total doses varied from about 50 to 1000 rads), and increases in thyroid cancer incidence were noted. The rather widespread practice of thymic irradiation, combined with low spontaneous frequency of malignancies of thyroid, facilitated observation of carcinogenic effects. It also gave rise to the impression that adult human thyroid was less sensitive to cancer induction than infant thyroid. Direct evidence on this point has, however, been equivocal.[32,36] For doses of a few hundred rads given in infancy, the likelihood of the development of a clinically detectable thyroid carcinoma is about 1% per 100 rads exposure. There is evidence of an increased level (about three times) of radiation-induced thyroid cancer in females compared to males and in Jews compared to non-Jews. For doses greater than 2000 to 2500 rads from external x rays, there is no clear association with induced cancer (probably reduced induction is due to cell killing).

It should also be pointed out that there was a greater likelihood of occurrence of benign rather than malignant nodules in these irradiated subjects. The largest group of children (about 3000 subjects and 5000 sibling controls) irradiated for *thymus enlargement* were followed for more than 25 years[27,37,38] and revealed 24 thyroid cancers and 52 thyroid adenomas in the irradiated group with no thyroid cancers and only 6 adenomas among controls. These studies also detected an increased risk of leukemia, as well as skin and salivary gland cancer.[39]

Children who received x irradiation to *tonsils* and *adenoids* have been reported to have shown an increased incidence of salivary and brain tumors. *Tinea capitis* patients (10,902 Jewish children), whose scalps were irradiated for treatment of the condition, have been reported to have an elevated risk of thyroid cancer even though the average dose to the thyroid was estimated to be only 6.5 rads. Since the level of risk per rad at this low dose is consistent with risk estimates obtained at doses of 100 rads or more for treatment of other benign conditions, these observations are consistent with a linear dose-incidence relationship.[28,40-41] Other high-dose regions (brain, skin, salivary gland, and bone marrow) have also been associated with cancer after these treatments.

In Utero X Irradiation

In a large retrospective study (the so-called Oxford survey), Stewart and associates[26,42-43] found an excess of cancer and leukemia deaths among children exposed in utero to diagnostic x rays. For an estimated mean dose of a few rads they demonstrated a marked increase (1.5 times spontaneous incidence) in the incidence of cancer and leukemia resulting in death before the age of ten. This relationship is strengthened by the observation that the likelihood of such malignancy appeared to increase linearly with the number of films.[43]

There is also evidence suggesting that exposure during the first trimester carries a higher relative risk (about 5 times higher) than the risk during the remaining parts of pregnancy.[1] These observations and the conclusions drawn from them are not without their critics. Others [1,42] have argued that the selection involves fetuses that may be more prone than the average to develop malignant disease spontaneously. There is also lack of support from animal experiments and from studies of the Japanese children exposed to A-bomb radiation in utero in which no excess cancer deaths were observed.[44] Mole[45] has explained this apparent lack of excess cancer deaths in the Japanese children on the basis that higher fetal doses from A-bomb exposure had cell-killing effects which reduced the subsequent incidence of cancer. However, a prospective study[45] in which medical indication did not play a part in the selection of the exposed subject (about 900 children exposed in utero on a routine basis were compared to 1300 children born before and after routine pelvic irradiation took place) also indicated no excess risk.

Mole[45] has pointed out that pelvic x-ray examinations are more frequently carried out (55% of cases) on mothers of twin pregnancies than on mothers of "singleton" pregnancies (10%). Despite this, the excess frequency of malignancy in twins irradiated in utero is no higher than that among irradiated singletons. The BEIR III[1] report states: "We consider the twin data to provide some of the strongest support for a causal relationship between in utero exposure to diagnostic x rays and the later increase in cancer risk."

In summary, an association between prenatal exposure to diagnostic x rays and an increased risk of developing malignancy during childhood has been established, but the question still remains of whether this represents a *causal* relationship and, if so, to what extent. It is possible that the radiation exposure involves a process of selection, resulting in an exposed population that differs from the unexposed population with regard to factors that influence the incidence of cancer. Such a selection process could account for part or all of the increased incidence of malignancy. Given the above, the risk of childhood cancer and leukemia from in utero exposure to low doses (0.2 to 20 rads) from diagnostic x rays is estimated by UNSCEAR, 1977[2] as being in the region of 200 to 250 cases per million exposed per rad, while the BEIR III estimate[1] is higher, 250 cases of leukemia and 300 cases of fatal cancer per million children exposed per rad.

Diagnostic X Ray

The literature contains reports which suggest that the relatively small doses of radiation involved in diagnostic x-ray studies performed on adults increase the incidence of cancer, particularly leukemia. Stewart, et al.[47] reported an association between the number of x rays and myelocytic leukemia, while a report by Gibson, et al.[48] showed an association, in men only, between leukemia and exposure to 21 or more x-ray films of the trunk. Both of these studies were conducted through questionnaires administered to patients and control subjects or to their next of kin, if the patient or control subject had died. Since little documentation of radiation exposure was possible through medical records, both studies have been seriously questioned.

More recently Bertell[49] has also suggested that the relatively small radiation doses from diagnostic x-ray procedures performed on adults increased the risk of myelocytic leukemia. A re-examination of results of these three reports[50] demonstrated that the available data did *not* support the conclusion that leukemia risk increased with the number of x-ray exposures. There is, however, a suggestion that a large number of diagnostic x-ray procedures (40 within a 10-year period) may accompany elevated leukemia risk, but the association does not necessarily demonstrate a cause-effect relationship.

A recent study[51] using centralized medical records allowed a comparison of the lifelong x-ray doses of 138 people who developed leukemia (all the leukemia cases reported in one county between 1954 and 1974) with matched controls. No significant increase in the risk of radiation-induced leukemia was found for doses less than 300 rads (when the doses were administered in small doses over long periods of time), leading to the conclusion that ordinary diagnostic x-ray examinations do not significantly increase risk of leukemia.

5.14 MEDICAL USE OF RADIONUCLIDES

Radium

Around the turn of the century there was fairly widespread use of radium medications, usually given internally in drinking water but also intravenously.[52] As late as 1931, a group of U.S. mental patients received weekly injections of ^{226}Ra (half-life of 1620 years). During the 1940s about 2000 German patients with bone tuberculosis and ankylosing spondylitis received re-

peated injections of ^{224}Ra (half-life of 3.6 days) as a treatment for their disease.[53] Osteosarcomas began to appear about 4 years after injection, peaked at 6 to 8 years, and returned to normal rates after 23 years. ^{224}Ra appears to be more effective in tumor induction than ^{226}Ra.

Thorium (^{232}Th)

Thirty to forty years ago thorium dioxide in colloidal form (called Thorotrast) was used as a contrast agent for arteriography or for visualizing liver and spleen.[54] The result of irradiation of liver, spleen, bone marrow, and endosteum has been the development of hemangiosarcomas and cholangiocarcinoma as well as acute leukemia.

Iodine (^{131}I)

There is currently no evidence, and it would probably be impossible to produce evidence, that any given *diagnostic radionuclide* test has produced malignant tumors. The associated radiation doses may be so low as to induce a number of cancers representing such a small fraction of the spontaneous cancer risk that it is unlikely that this increment (if indeed there is one) could be distinguished from the background.

The best evidence of how small the risk is can be seen in the recent report of Hohn, Lundell, and Wallinder.[55] This study measured the incidence of malignant thyroid tumors in humans after exposure to *diagnostic* doses of ^{131}I. This was a particularly informative study since thyroid is sensitive to radiation-induced cancer, and the radiation dose delivered by ^{131}I to the thyroid gland is relatively high even for diagnostic studies (about 1 rad/μCi administered). In this study 10,216 adult Swedish patients, from 1952 to 1965, were given an average dose of 60 μCi of ^{131}I, with an average thyroid uptake of 24 μCi. Nine malignant thyroid tumors were found in the treated group, while the number of expected spontaneous thyroid tumors computed from the

Swedish cancer-incidence figures was 8.3. The authors concluded that there was no elevation of the incidence of malignant thyroid tumors in the patients who had received diagnostic doses of [131]I.

Treatment of hyperthyroidism with radioactive iodine has been investigated for its carcinogenic potential (principally leukemia and thyroid cancer). Sporadic reports in the 1950s of leukemia in patients treated with radioiodine led to two large-scale studies. In 1960, Pochin[56] analyzed data involving over 59,000 patients and concluded that there was no evidence of leukemia induction by [131]I. In 1968, results of a large-scale investigation (the Cooperative Thyrotoxicosis Study) involving 18,400 patients treated with ^{131}I and 10,700 patients treated with surgery were reported.[57] The study showed no increased risk of leukemia. The average follow-up, however, was only eight years. The incidence of leukemia in patients treated with radioiodine was no different from that for surgically treated patients, but for some reason, *both* groups had a higher incidence of leukemia than the population at large (this points to the importance of an appropriate group of control subjects). If thyroid cancers are being induced by ^{131}I treatment for thyrotoxicosis, the incidence is apparently quite low. The data from the Cooperative Thyrotoxicosis Follow-up Study[57] showed no serious risk of thyroid carcinoma (average follow-up, however, of only eight years). On the other hand, the Marshall Islands natives[31,32] accidentally exposed to ^{131}I from radioactive fallout developed both benign and malignant neoplasms,[31,32] but their thyroid glands received *less* radiation than those of patients treated for hyperthyroidism.

Treatment of thyroid cancer with ^{131}I results in high doses in bone marrow (100 mCi dose of ^{131}I results in a red-marrow dose of about 30 rads). Increased risk of malignancy from these high doses has been reported.[58,59] In some 200 patients treated for thyroid cancer with radioiodine, it was observed that 12 malignancies occurred, although only 5.2 were expected in a sample of this size. Of these, four were acute leukemia (less than one expected), and four were carcinoma of the breast (one expected). All of these differences were highly significant.

Phosphorus (^{32}P)

It has been known since 1945 that acute leukemia occurs in patients treated with ^{32}P, but acute leukemia also occurs in patients with polycythemia vera treated only with drugs or phlebotomy.[58] It is possible that ^{32}P, by prolonging life, permits the natural evolution of polycythemia vera into acute leukemia, or that the leukemogenic action of ^{32}P may result from irradiation of a sensitive population.

5.15 OCCUPATIONAL EXPOSURE

Radiologists

The first neoplasms attributed to radiation injury were *skin cancers*, mainly on the hands and forearms of radiologists and their assistants exposed repeatedly through the use of primitive equipment in earlier times. Within 15 years after Roentgen's discovery of x rays, almost 100 cases of skin cancer had been reported in America and Europe (half in physicians, half in technicians). Although chronic radiodermatitis was long thought to be a prerequisite for induced skin cancer, it is now amply documented that basal-cell cancer, in particular, can occur in skin with little or no evidence of radiation damage.[1] It has been suggested that the likelihood of radiogenic skin cancer is much greater in the presence of prior radiodermatitis, but the controversy is not completely resolved.

Leukemia came to be acknowledged in the 1940s to occur in increased frequency among radiologists. A number of published reports on causes of death covering time intervals in the 1920s and 1930s compared radiologists (as determined from death notices in the *Journal of the American*

Medical Association) with other physicians and the general population. Most of these reports indicated a significant excess of leukemia deaths in radiologists (some eight to ten times higher than other physicians). A widely quoted study by Lewis[60] surveyed 425 death certificates of radiologists who died between the ages of 35 and 74 during the years 1948 to 1961. It revealed 12 cases of leukemia as compared to 4 expected cases. Furthermore, all the cases of leukemia in radiologists were of the types generally induced by radiation—there were no cases of chronic lymphoid leukemia, the leukemia type least likely to be induced by irradiation.

Perhaps the best data are those of Seltser and Sartwell[61] who compared deaths from leukemia from 1935 to 1958 between members of the Radiological Society of North America (RSNA) and the American Academy of Ophthalmology and Otolaryngology (AAOO). Using a life table method of analysis, they demonstrated a 2.5 times greater mortality from leukemia in radiologists. They also demonstrated a 1.6 times increase in the incidence of all other cancers as well as a reduction of 4.8 years in the median life span when members of the RSNA were compared to the AAOO for the period 1935 through 1948. This is almost identical to the difference in the mean age at death (5.2 years) obtained by Shields Warren[62,63] when he compared age at death of radiologists with age at death of physicians who did not routinely use radiation (from 1930 through 1954). Little is known about the doses received by early radiologists, but there is general agreement that they were probably high. For those who died in the 1930s and 1940s, the possible lifetime (40-year) occupational exposure has been estimated to be between 800 and 2000 rads.

British radiologists entering the profession before 1921 suffered a death rate from cancer and leukemia 75% higher than that of medical practitioners, with no significant increase after this time.[64]

Radium Dial Painters

The radium dial painters were mainly young women who ingested large quantities of radium in the 1920s as a consequence of tipping their brushes with their lips as they applied luminous paint, containing radium, to watch dials. This material, once ingested, was deposited in their bones. These women had an excess risk of osteogenic sarcomas of the bones in the head and neck area as well as cancers of the paranasal sinuses.[52]

Miners

Workers in uranium mines are exposed to high levels of alpha radiation from radium and its daughters carried on inhaled dust. Epidemiologic studies have indicated a dose-response relationship between exposure to airborne radiation and the incidence of lung cancer.[28] Others mining fluorspar, tungsten, iron, and lead who are also exposed to internal alpha radiation, have been shown to have elevated lung cancer risks as well.

SUMMARY

1. Ionizing radiation is a *general* carcinogen capable of inducing tumors in almost all tissues of the body.
2. The exact mechanism by which radiation transforms normal cells to cancerous cells is unknown, but most theories center around the concept of *somatic mutation*.
3. There is usually a latent period of a few to tens of years between irradiation and appearance of cancer.
4. There is little direct information relating increased incidence of cancer in populations exposed to doses much below 50 to 100 rads.
5. The expected incidence of cancer resulting from doses lower than 50 to 100 rads must usually be estimated by interpolation from observed inci-

dence resulting from exposure to higher doses. A linear extrapolation from higher to lower doses is commonly used, but there is much evidence to indicate that this approach may *overestimate* the actual incidence. The linear-quadratic model has been used recently by the BEIR III Committee as its preferred dose-response relationship.

6. Induction of cancer is the most important somatic effect of low-level radiation.

7. The major radiation-induced cancers are female breast, thyroid, lung, and leukemia.

8. Estimates of total radiation-induced cancer mortality center around 100 lifetime cases per rad in a population of 1 million, as compared to a normal risk of about 160,000 lifetime fatal cancers per million.

9. Most of the information relating human radiation exposure to induced malignancy comes from studies of four major groups:

   a. Japanese A-bomb survivors—(increased incidence of leukemia over a wide dose range).

   b. Medically irradiated patients— (ankylosing spondylitis and leukemia; thymic irradiation and thyroid cancer; diagnostic and therapeutic irradiation and breast cancer; radiation castration and leukemia in females; and cancer and leukemia following in utero irradiation).

   c. Patients who developed subsequent cancer and leukemia from internally deposited radioactive materials (radium, thorium, ^{32}P, and ^{131}I).

   d. Early radiologists who developed cancer and leukemia.

REFERENCES

1. National Research Council, Advisory Committee on the Biological Effects of Ionizing Radiation (BEIR III): The Effect on Populations of Exposure to Low Levels of Ionizing Radiation. Washington, D.C., National Academy of Science, 1980.
2. United Nations Report of the United Nations, Scientific Committee on the Effects of Atomic Radiation (UNSCEAR). Official Records of the General Assembly, Thirty-Second Session, Supplement No. 40 (A/32/40). New York, 1977.
3. Beebe, G.W.: What knowledge is considered certain regarding human somatic effects of ionizing radiation. Issue Paper No. 1. Report to the Subcommittee to Develop Federal Strategy for Research into the Biological Effects of Ionizing Radiation. Presented at a public meeting at the National Institutes of Health, Bethesda, MD, March 10, 1980.
4. Crump, K.S., et al.: Fundamental carcinogen processes and their implications for low dose risk assessment. Cancer Res., 36:2973–2979, 1976.
5. Rossi, H.H.: Interrelation between physical and biological effects of small radiation doses. In Biological and Environmental Effects of Low-Level Radiation. Vol. 1. Vienna, International Atomic Energy Agency, 1976, pp. 245–251.
6. Beebe, G.W., Kato, H., and Land, C.E.: Studies of the mortality of A-bomb survivors. Mortality and radiation dose, 1950–1974. Radiat. Res., 75:138–201, 1978.
7. Brown, R.F.: Diagnostic procedures. Science Projection Paper A. Report to the Subcommittee to Develop Federal Strategy for Research into the Biological Effects of Ionizing Radiation. Presented at a public meeting at the National Institutes of Health, Bethesda, Md., March 10, 1980.
8. Land, C.E.: Estimating cancer risks from low doses of ionizing radiation. Science, 209:1197–1203, 1980.
9. Sinclair, W.K.: Effects of low-level radiation and comparative risk. Radiology, 138:1–9, 1981.
10. Upton, A.C., Randolph, M.L., and Conklin, J.W.: Late effects of fast neutrons and gamma rays in mice as influenced by the dose rate of irradiation: induction of neoplasia. Radiat. Res., 41:467–491, 1970.
11. Ulrich, R.L., and Storer, J.B.: The influence of dose, dose rate and radiation quality on radiation carcinogenesis and lifeshortening in RFM and BALB/c mice. In Late Biological Effects of Ionizing Radiation. Vol. 2. Vienna, International Atomic Energy Agency, 1978, pp. 95–113.
12. Brown, J.M.: Linearity vs non-linearity of dose response for radiation carcinogenesis. Health Physics, 31:231–245, 1976.
13. NCRP Report No. 64: Influence of Dose and its Distribution in Time on Dose-Response Relationships for Low-LET Radiations. Washington, D.C., National Council on Radiation Protection and Measurements, 1980.
14. Kellerer, A.M., and Rossi, H.H.: RBE and the primary mechanism of radiation action. Radiat. Res., 47:15–34, 1971.
15. Kellerer, A.M., and Rossi, H.H.: The theory of dual radiation action. Current Topics Radiat. Res. Q., 8:85–158, 1972.
16. Cohen, B.L.: The cancer risk for low-level radiation. Health Physics, 39:659–678, 1980.

17. Mancuso, T.F., Stewart, A., and Kneale, G.: Radiation exposures of Hanford workers dying from cancer and other causes. Health Physics, 33:369–385, 1977.

18. Hutchinson, G.B., MacMahon, B., Jablon, S., and Land, C.E.: Review of report by Mancuso, Stewart, and Kneale of radiation exposure of Hanford workers. Health Physics, 37:207–220, 1979.

19. Bross, I.D.J., and Natarajan, N.: Genetic damage from diagnostic radiation. JAMA, 237:2399–2401, 1977.

20. Oppenheim, B.E.: Genetic damage from diagnostic radiation? A critique of the Bross and Natarajan study. JAMA, 242:1390, 1979.

21. Najarian, T., and Colton, T.: Mortality from leukemia and cancer in shipyard nuclear workers. Lancet, 1:1018–1020, 1978.

22. Caldwell, G.G., Kelley, D.B., and Heath, C.W.: Leukemia among participants in military maneuvers at a nuclear bomb test. A preliminary report. JAMA, 244:1575–1578, 1980.

23. Interagency Task Force on the Health Effects of Ionizing Radiation: Report of the Work Group on Science. Washington, D.C., U.S. Department of Health, Education, and Welfare, 1979.

24. Frigerio, N.A., and Stowe, R.S.: Carcinogenic and genetic hazard from background radiation. IAEA Conference on Low-Level Radiation, Chicago, Ill., 1976. (IAEA/SM-202/805), pp. 385–392.

25. ICRP 26. Recommendations of the International Commission on Radiological Protection. Annals of the ICRP, Vol. 1, No. 3, International Commission on Radiological Protection. New York, Pergamon Press, 1977.

26. Stewart, A., and Kneale, G.W.: Radiation dose effects in relation to obstetric x-rays and childhood cancer. Lancet, 1:1185–1187, 1970.

27. Hempleman, L.H., et al.: Neoplasms in persons treated with x rays in infancy: Fourth survey in 20 years. J. Natl. Cancer Inst., 55:519–530, 1975.

28. Modan, B., Ron, E., and Werner, A.: Thyroid cancer following scalp irradiation. Radiology, 123:741–744, 1977.

29. Band, P., et al.: Potentiation of cigarette smoking and radiation. Evidence from a sputum cytology survey among uranium miners and controls. Cancer, 45:1273–1277, 1980.

30. Finch, S.C.: The study of atomic bomb survivors in Japan. Am. J. Med., 66:899–901, 1979.

31. Dobson, R.L., and Straume, T.: Letter to the Editor. Science, 213:8, 3 July 1981.

32. Maxon, H.A., et al.: Ionizing irradiation and the induction of clinically significant disease in the human thyroid gland. Am. J. Med., 63:967–978, 1977.

33. Court Brown, W.M., and Doll, R.: Mortality from cancer and other causes after radiotherapy for ankylosing spondylitis. Br. Med. J., 2:1327–1332, 1965.

34. Boice, J.D., et al.: Risk of breast cancer following low-dose radiation exposure. Radiology, 131:589–597, 1979.

35. Smith, P.G., and Doll, R.: Late effects of x-irradiation in patients treated for metropathia haemorrhagica. Br. J. Radiol., 49:224–232, 1976.

36. See reference 1, page 302.

37. Favus, M.J., et al.: Thyroid cancer occurring as a late consequence of head-and-neck irradiation. N. Engl. J. Med., 294:1019–1025, 1976.

38. Shore, R.E., et al.: Radiation and host factors in human thyroid tumors following thymus irradiation. Health Physics, 38:451–465, 1980.

39. Interagency Task Force on the Health Effects of Ionizing Radiation, The Libassi Report. Washington, D.C., U.S. Department of Health, Education and Welfare, 1979, pp. 31–38.

40. Carrol, R.G.: The relationship of head and neck irradiation to the subsequent development of thyroid neoplasms. Semin. Nucl. Med., 6:411–424, 1976.

41. Harley, N.H., et al.: Follow-up study of patients treated by x-ray radiation for tinea capitis. Estimation of the dose to the thyroid and pituitary glands and other structures of the head and neck. Phys. Med. Biol., 21:631–642, 1976.

42. Totter, J.R., and MacPherson, H.G.: Do childhood cancers result from prenatal x rays? Health Physics, 40:511–524, 1981.

43. Stewart, A.M., and Kneale, G.W.: Radiation dose effects in relation to obstetric x rays and childhood cancers. Lancet, 2:1185–1188, 1970.

44. Jablon, S., and Kato, H.: Childhood cancer in relation to prenatal exposure to atomic-bomb radiation. Lancet, 2:1000–1003, 1970.

45. Mole, R.H.: Ionizing radiation as a carcinogen: Practical questions and academic pursuits. Br. J. Radiol., 48:157–169, 1975.

46. Oppenheim, B.E., Greim, M.L., and Meier, P.: Effects of low-dose prenatal irradiation in humans. Analysis of Chicago lying-in data and comparison with other studies. Radiat. Res., 57:508–544, 1974.

47. Stewart, A., Pennybacker, W., and Barber, R.: Adult leukemias and diagnostic x-ray. Br. Med. J., 2:882–890, 1962.

48. Gibson, R., et al.: Irradiation in the epidemiology of leukemia among adults. J. Natl. Cancer Inst., 48:301–311, 1972.

49. Bertell, R.: X-ray exposure and premature aging. J. Surg. Oncol., 9:379–391, 1977.

50. Ginevan, M.E.: Nonlymphatic leukemias and adult exposure to diagnostic x-rays: the evidence revisited. Health Physics, 38:129–138, 1980.

51. Linos, A., et al.: Low-dose radiation and leukemia. N. Engl. J. Med., 302:1101–1105, 1980.

52. Rowland, R.E.: Dose and damage in long-term radium cases. In Medical Radionuclides: Radiation Dose and Effects. Edited by R.J. Cloutier, C.L. Edwards, W.S. Snyder, and E.B. Anderson. USAEC CONF-691212, Clearinghouse for Federal Scientific and Technical Information, National Bureau of Standards, U.S. Department of Commerce, Springfield, Va., 1970, pp. 369–386.

53. Speiss, H., and Mays, C.W.: Bone cancers induced by ^{224}Ra (ThX) in children and adults. Health Physics, 19:713–729, 1970.

54. de Silva Borta, J., da Motta, L.C., and Tavares, M.H.: Thorium dioxide effects in man. Environ. Res., 8:131–159, 1974.

55. Hohn, L.E., Lundell, G., and Wallinder, G.: In-

cidence of malignant thyroid tumors in humans after exposure to diagnostic doses of iodine 131. Retrospective study. J. Natl. Cancer Inst., 64:1055–1059, 1980.

56. Pochin, E.E.: Leukemia following radioiodine treatment of thyrotoxicosis. Br. Med. J., 2:1545–1550, 1960.
57. Saenger, E.L., Thoma, G.E., and Tompkins, E.A.: Incidence of leukemia following treatment of hyperthyroidism. Preliminary report of the Cooperative Thyrotoxicosis Therapy Follow-up Study. JAMA, 205:855–862, 1968.
58. Berlin, N.I., and Wasserman, L.R.: The association between systemically administered radioisotopes and subsequent malignant disease. Cancer, 37:1097–1101, 1976.
59. Pochin, E.E.: Long term hazards of radioisotope treatment of thyroid cancer. *In* Thyroid Cancer.

UICC Monograph Series, Vol. 12. Berlin, Springer, 1969, pp. 293–304.

60. Lewis, E.B.: Leukemia, multiple myeloma, and aplastic anemia in American radiologists. Science, 142:1492, 1963.
61. Seltser, R., and Sartwell, P.E.: The influence of occupational exposure to radiation on the mortality of American radiologists and other medical specialists. Am. J. Epidemiol., 81:2–22, 1965.
62. Warren, S.: Longevity and causes of death from irradiation in physicians. JAMA, 162:464–468, 1956.
63. Warren, S.: Radiation carcinogenesis. Bull. NY Acad. Med., 46:131–147, 1970.
64. Smith, P.G., and Doll, R.: Mortality from cancer and all causes among British radiologists. Br. J. Radiol., 54:187–194, 1981.

Chapter 6

Effects of Radiation on Embryonic and Fetal Development

6.1 INTRODUCTION

Previous chapters have focused on effects expected when radiation dose is so low that cellular damage is for the most part sublethal. While it is true that small radiation doses produce *principally* sublethal damage, some cells are killed as well. In adults, the loss of a few cells is probably an inconsequential matter. In all likelihood the deficit will either not be felt at all or will be quickly made up.

In the developing embryo and fetus, however, the loss of small numbers of cells is a different matter. These stages in the lives of living beings are composed of relatively small numbers of cells, each of which is to be ancestral to a great many cells in the postembryonic body. If some are killed, they are not so easily replaced, and the descendants they would have produced may be missing. The result can be a postembryonic organism that is smaller than average, one that is missing some of its parts, or one with parts that are poorly or inadequately developed. Even if only a few cells are killed during early embryonic development, too few may remain to produce a viable organism, and intrauterine

death may result. Later in development, after the embryo has grown, it may be composed of enough cells so that the loss of a few is unlikely to be lethal. However, the organs may then be composed of so few cells that the loss of some of them can result in deformities. It can also result in the production of smaller-than normal babies at birth.

In addition to the far-reaching consequences of cell loss during embryonic and fetal development, embryonic tissue is expected to respond more to given radiation doses than that of adults. That is, *given* radiation doses probably kill more cells in embryonic tissue than in adult tissue. There are several reasons for this—embryonic cells are differentiating, and *differentiating* cells are more radiosensitive than *differentiated* ones; and embryos have high rates of mitotic activity, and proliferating cells are more radiosensitive than quiescent ones. In short, comparable low radiation doses probably kill more embryonic cells than adult cells, and this combines with the fact that the loss of small numbers of cells has more far-reaching consequences for embryos and fetuses than for adults to make prenatal life exceptionally vulnerable to irradiation.

The result of cell killing in embryonic or fetal life is often referred to as the "classic triad" of radiation-induced embryonic effects[1] and includes intrauterine and/or postnatal growth retardation; embryonic, fetal, or perinatal death; and congenital malformations.

The probability of finding one or more of these effects in an irradiated embryo is dependent upon a number of factors, including radiation dose, dose rate, and stage of gestation or development at which exposure occurred.

6.2 STAGE OF DEVELOPMENT

Ionizing radiation is not the only agent that kills or deforms developing embryos or changes their developmental potential, nor are the effects it produces characteristic. Many agents (chemicals, viruses, and other physical agents) cause abnormal embryonic development, and the abnormalities produced are identical with those caused by exposure to ionizing radiation. Abnormalities of development are more characteristic of and dependent on the organ system or body part developing *at the time* at which embryo-deforming agents are used than upon the agents themselves. Differentiating, proliferating cells are killed, and resulting cell loss is expressed as incomplete or abnormal development of the system in which it happens. Radiation interactions are random and indiscriminate. Therefore, when entire embryos are irradiated, different degrees of response will be observed among their organ systems depending on which of them is developing at the time of irradiation. The *stage of development* of the embryo, then, determines which organ system of totally irradiated embryos will be most seriously damaged and, consequently, the kind of abnormality or abnormalities (if any) that will be observed in postnatal life.

Based principally on rodent experiments, good information about sensitivity of mammalian embryos at various stages in development has been collected. The pattern observed is given in Fig. 6.1.

These data have been extrapolated from rodent data to predict results of irradiation of human embryos at comparable gestation times. Figure 6.1 is based upon the concept that, although the rodent gestation period is much shorter than that of humans (21 vs 266 days), cleavage, implantation, placentation, organogenesis, and refined differentiation occur in all mammals in the same sequence. Thus, if a human embryo is subjected to irradiation at a development stage *comparable* to that of the mouse, similar damage should result. An examination of Fig. 6.1 shows that exposure of the embryo *during the preimplantation stage* results in a high incidence of prenatal death. This probably relates to the fact that the embryo then consists of very few cells, and the loss of even a single cell has some chance of being fatal. Congenital malformations rarely result from irradiation during the preimplantation stage.

The type of embryonic damage changes rather abruptly with the onset of organogenesis. The incidence of malformation rises dramatically at this time and is elevated throughout early organogenesis. *All* organ systems begin to form then, but differentiation of cells to form particular or-

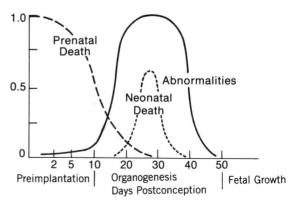

Fig. 6.1. Effects of in utero exposure on different gestational days on the induction of lethality and major abnormalities in the embryo. (Redrawn from Russell, L.J. and Russell, W.L.[2])

gans begins on specific days, resulting in specific abnormalities. The incidence of prenatal death is much reduced following irradiation during major organogenesis; however, there is an increase in perinatal death (death at or around the time of birth), particularly after higher doses. This is at least partly due to the presence of severe abnormalities, caused during early organogenesis, which are fatal at term. The high incidence of gross malformations falls off rapidly as organogenesis progresses.

Irradiation during early organogenesis and, even more so during late organogenesis, probably produces an irrecuperable loss of cells following higher doses, which results in significant growth retardation.

6.3 LETHALITY (INTRAUTERINE DEATH)

Embryonic death is a component of the triad of radiation effects observed in irradiated mammalian embryos and is considered here first, because of its prominence in the preimplantation stage of pregnancy. In rats and mice lethal effects of radiation are greatest on the first day postconception and decline thereafter, gradually approaching adult sensitivity at term (Table 6.1). As pregnancy continues, radiation doses that are lethal become higher and

higher until, at term, lethal doses are about the same as those for adults. Doses *as low as 10 rads* given to *early*, preimplantation mouse embryos, have been reported to reduce litter size.[3]

6.4 MALFORMATIONS

Malformations resulting from irradiation of mouse and rat embryos in the preimplantation period are not commonly observed, regardless of dose. Most reported exceptions come from animal strains which have an elevated *spontaneous* incidence of specific abnormalities. For instance, Rugh[4] reported increased incidence of exencephaly in preimplantation mouse embryos given 15 to 25 R, but the mouse strain used exhibits a 1% spontaneous incidence of this anomaly.

In rats and mice, irradiation is most effective at producing malformations during the early period of organogenesis. In rodents this occurs between the sixth and fourteenth day postconception and is approximately equivalent to the twelfth to fiftieth day of human pregnancy. Gross malformations have been produced at low frequency in rats and mice using doses in the range of 25 to 50 rads delivered during early organogenesis. As dose is increased above 50 rads the likelihood of malforma-

Table 6.1. *Estimates of minimum effective doses of radiation for the human embryo based on compilation of mouse, rat, and human data. From Brent and Gorson, 1972*[1]

Age	Approximate Minimum Lethal Dose	Minimum Dose for Non-recuperable Growth Retardation in Adults	Minimum Dose for Recognizable Gross Malformations
Day 1	10 R*	a	b
Day 14	25 R	25 R	—
Day 18	50 R	25–50 R	25 R
Day 28	50 R	50 R	25 R
Day 50	100 R	50 R	50 R
Late Fetus to Term		50 R	50 R

a) Surviving embryos are not growth retarded even after high doses of radiation.
b) Malformation incidence is extremely low even after high doses of radiation.

*R = roentgens

tions increases. For example, a dose of 200 rads given to mouse embryos during that period can result in 100% incidence of malformation at birth.

The most common radiation-induced developmental anomalies are central nervous system damage (microcephaly and associated mental retardation being most commonly observed), sense organ damage, and stunted growth. The production of specific malformations is associated with irradiation during a *definite time* in the period of organogenesis, usually during the appearance of the *first* morphologic evidence of differentiation in the organ or portion of the organ involved. Reduced dose rate can have a profound effect. Long-term studies in the mouse[3] and rat[5] have demonstrated that, approximately 2 rads per day for 10 or 11 generations respectively, did not alter incidence of malformation, stillbirths, litter size, or fertility.

It is very difficult to identify radiation as a proved cause of human abnormalities after exposure to *low* doses for two reasons—radiation produces no unique abnormalities, and the incidence of spontaneous malformations in the population is approximately 4 to 6%.[1] This large background makes it difficult to convincingly demonstrate small increases in frequency—if indeed they occur. Thus, the documented evidence of the effects of radiation on developing human embryos is data taken from atomic-bomb survivors and from therapeutically exposed pregnant women.[6] Both groups received high levels of radiation (the atomic-bomb survivors, up to 200 rads; most therapeutic exposures, above 200 rads). The predominant types of malformations observed were of the central nervous system (microcephaly, mental retardation), sense organs, and skeletal system. The similarity of these findings to those obtained at the same high dose levels in irradiated rodents lends credence to the concept that embryonic damage similar to that observed in embryonic rodents irradiated at lower doses will also be produced

in humans. Thus, the possibility of *major* malformations resulting from radiation doses in the diagnostic range (less than 10 rads) appears to be very small. *Major* malformations involve the absence of or gross distortion of a tissue, organ, or tissue component, or the presence of easily observable tissue hypoplasia such as microcephaly, microphthalmia, testicular atrophy, or cerebral hypoplasia. For instance, human epidemiological studies that compared incidence of malformations showed no difference between those exposed in utero to diagnostic x rays and those not exposed.[7]

6.5 GROWTH RETARDATION

Given high enough radiation doses, intrauterine growth retardation can result from radiation delivered anytime during gestation *after implantation*. The best evidence shows that irradiated *preimplantation* embryos that survive to term grow normally in the pre- and postpartum periods. The implanted embryo, however, is susceptible to growth retardation from radiation exposure even when it is not susceptible to malformation—that is, after major organogenesis. Animal experiments indicate that embryos exposed to 100 rads or more after implantation exhibit some growth retardation which may persist into adulthood.

A threshold for growth retardation seems to exist and appears to lie below 100 rads (but still well above the diagnostic range) and depends upon the stage of gestation and the dose rate. Brent and Gorson[1] have demonstrated that in mice, the lowest dose that produces growth retardation is in the range of 30 to 60 rads. There is also substantial evidence to indicate that doses *greater* than 100 rem contributed to growth retardation in children irradiated in utero both at Hiroshima and Nagasaki.[8] This was in the form of reduced head circumference and reduced height and weight. Finally, the literature indicates that animal embryos irradiated at any stage of gestation with less than 25 rads have not exhibited growth retardation; at least they fall into the normal range of weights and sizes as adults.

Table 6.2 presents in summary form the probability of occurrence of the members of the classic triad as a function of stage of gestation or time postconception. It is well to recall that the effects listed are not necessarily restricted to the corresponding time periods, but that the effects listed *are most likely* to occur during those time intervals, *if* extrapolation from the abundant lower animal data and the application of the scanty human data lead to accurate predictions. It does not mean that any of the effects listed *inevitably* occur. Several human fetuses have been given very high doses of radiation for one reason or another and have been born alive without evidence of malformation and within average or expected ranges of weight and size. Furthermore, it does not mean that effects *not* listed with particular postconception times certainly will not occur. Intrauterine death, for example, can occur if irradiation is performed between 13 and 50 days postconception, but it is an unlikely occurrence, especially after low doses. All that is meant is that *if* irradiation is to cause any adverse effect during in utero life, the effects listed are those *most likely* to occur.

6.6 STERILITY AND GENETIC EFFECTS

Although the "classic triad" represents the effects most commonly associated with intrauterine irradiation, other pathologic effects have been observed in adult organisms after in utero irradiation. Some of these, including production of mutations and sterility, are not apparent at birth, and may be seen only in later life or in later generations. It is quite difficult to present general patterns of response of *embryonic* male and female gonads to irradiation. There is considerable variation among species and between sexes and in sensitivity at various stages of gestation. However, from data available on both humans and animals, one may tentatively conclude that acute embryonic doses less than 25 rads will not result in sterility in either human males or females.

Although there are few published reports directly addressing the effect of embryonic irradiation on mutation rate, there is little evidence to indicate that embryonic gonads are more sensitive to mutation than those of adults.

Table 6.2. *Effects of irradiation on the human embryo*

Days After Fertilization	Period of Development	Effects
1–9	Preimplantation	Most probable effect is death. Little chance of malformation.
10–12	Implantation	Reduced probability of lethality. Malformation still unlikely. Intrauterine growth retardation is predominant effect.
13–50	Early organogenesis	Production of malformation. Retarded growth.
51–Term	Late organogenesis and fetal growth	Effects on central nervous system. Growth retardation with higher doses.
All		Possibility of increased incidence of leukemia.

6.7 APPLICATION TO MEDICAL PRACTICE

The purpose of the remainder of this chapter is to apply the basic information just presented to medical practice, keeping in mind the limitations in our knowledge.

6.8 RADIATION DOSE TO HUMAN EMBRYOS DURING MEDICAL PROCEDURES

The use of the term "dose estimate" in place of embryonic "dose" is well advised when discussing diagnostic radiologic and nuclear medical procedures (Tables 6.3 and 6.4). Doses in nuclear medicine are generally based upon a variety of assumptions. Standard anatomic models are used; and uniform distribution of radioactivity, "average" biologic distribution, "average" excretion routes and "average" rate of excretion for radiopharmaceuticals are all assumed. There is no certainty, however, that all or any of these apply precisely in any given case. In particular, distribution of radioactivity and excretion of radiopharmaceuticals may well deviate from these "averages" in sick people.

Embryonic doses associated with diagnostic radiologic procedures probably also vary considerably because of the variety of techniques used to perform the same study. There may be varying numbers of radio-

Table 6.4. *Dose estimates for human embryo—x-ray examinations*

Examples of Procedures Yielding	
High Doses: (1–2 rad)	Barium enema (with fluoroscopy)
Examples of Procedures Yielding	
Medium Doses: (0.5–1.0 rad)	Pelvimetry, upper GI series (with fluoroscopy), IV or retrograde pyelogram, lumbar and lumbosacral spine.
Examples of Procedures Yielding	
Low Doses: (0.1–0.5 rad)	Cholecystography, cholangiography, pelvis, hip, abdomen (KUB).
Examples of Procedures Yielding	
Very Low Doses: (<0.1 rad)	Upper and lower extremities, skull, chest, cervical and thoracic spine.

graphs, varying amounts of fluoroscopy time, different kilovoltage and/or different screen-film combinations used by different practitioners or under different circumstances to perform the same procedure. Embryonic dose estimates from nuclear medical procedures rely entirely on calculated values, but in diagnostic radiology the dose can be estimated more closely by dose measurements in a body-equivalent "phantom." It is apparent from Tables 6.3 and 6.4 that from most *common* diagnostic procedures using x rays or radiopharmaceuticals, radiation dose to embryo or fetus would be less than one rad.

Table 6.3. *Dose estimates for human embryo—nuclear medicine*[9]

Radiopharmaceutical	Amount Administered	Dose (Rad)
^{99m}Tc-Sodium pertechnetate	15 mCi	0.56
^{99m}Tc-Sulfur colloid	2 mCi	0.014
^{99m}Tc-Polyphosphate	15 mCi	0.30
^{131}I-Sodium iodide	0.1 mCi	0.01*
^{131}I-Sodium rose bengal	0.15 mCi	0.10
^{99m}Tc-Albumin	1 mCi	0.01

*Prior to ten weeks gestation when fetal thyroid function begins.

6.9 RADIOIODINE AND FETAL THYROID

Fetal thyroid takes up iodine by the tenth week of gestation[10] either to the same degree as or to a greater degree than maternal thyroid. Although inorganic iodine readily crosses placenta, iodine attached to proteins, rose bengal, and other substances is less likely to do so. However, iodide ions may be released from these and other radiopharmaceuticals both on the shelf and in the body and can cross placenta. Incorporation of iodine in molecules which are excluded by placenta does not, therefore, absolutely guarantee no uptake of iodine in fetal thyroid.

Pathologic effects including destruction of fetal thyroid[11,12] have been reported when therapeutic doses of ^{131}I have been administered to pregnant women. Tracer or diagnostic doses of radioactive iodine have not been reported to produce deleterious effects in the fetus, but there remains the *possibility* of inducing thyroid cancer in susceptible individuals by prenatal irradiation of thyroid, even from small quantities of radioiodine.

If radioactive iodine is to be administered for *therapeutic* purposes to potentially pregnant patients, it is advisable to carry out a test of pregnancy prior to administration.[11] Though falsely negative tests are possible early in pregnancy, false negatives are very unlikely by the tenth week of gestation, the putative time of onset of thyroid function.

6.10   ELECTIVE BOOKING; THE "10- OR 14-DAY RULE"

In 1970, the NCRP[13] recommended that:

"When radiologic procedures are planned on pregnant or potentially pregnant women . . . it is recommended that radiologic examinations of the abdomen and pelvis which do not contribute to the diagnosis or treatment of such women in relation to their current illness (e.g., chronic low-back pain; x-ray examination for employment) be restricted to the first 14 days of the menstrual cycle in the case of potentially pregnant individuals and avoided entirely during known pregnancy . . . Examinations of the abdomen and pelvis that are deemed useful to patient care may be done at any time without regard for the phase of the menstrual cycle or fetal presence. In each case, the final decision to proceed or not to proceed must reside with the attending physician . . ."

Some have strongly supported this concept and extended it to *nonelective* examinations as the so-called "10- or 14-day rule." Specifically, they recommend that whenever

possible, pelvic irradiation of all women of childbearing age be restricted to the 10 or 14 days following the onset of menses. This reduces chances of embryonic irradiation during the period of organogenesis, since there is relatively little chance of unsuspected pregnancy during this time. There usually are a number of problems in implementing such a program and the following are reasons to question its wisdom.[1]

1. The logistics of scheduling "potentially pregnant" women only in the early days of the menstrual cycle can be very complex.
2. The literature indicates that deleterious effects of radiation in unborn children are possible at all stages of pregnancy (see Table 6.2).
3. In a potentially pregnant patient with new symptoms, the decision to wait to do a radiologic examination until the beginning of the next menstrual cycle may in fact be a decision to wait at least three months (or possibly the full term of pregnancy) before evaluating these symptoms—if one is trying to avoid irradiation of the embryo, particularly in the first trimester. The likely deleterious result of irradiation during the first two weeks is resorption of the preimplantation embryo. Postponement of the procedure confronts physicians and patients with a decision, should pregnancy be established, of whether to conduct the examination during major organogenesis, or to delay it until the fetal growth stage or postpartum.
4. Since cytogenetic (chromosomal) and genetic effects are thought to *diminish* with time after radiation exposure, the concentration of large numbers of radiologic examinations in the first half of the menstrual cycle might have a deleterious effect. Brent and Gorson[1] have stated: "No one at this time is able to determine which hazard is more deleterious to the human race—

a dose of less than 10 rads to the ovaries of every woman exposed *just* before a small percentage of them conceived or less than 10 rads to the predifferentiated human zygotes of the small percentage of women who are pregnant'' (italics added).

6.11 RECOMMENDATIONS FOR POTENTIALLY PREGNANT PATIENTS

Based upon the previous discussion the following recommendations[14] are made:

1. Avoid radiologic procedures not essential for optimal medical care.
2. Ask referring physicians to indicate on the consultation request if the patient is pregnant or potentially pregnant (date of last menstrual period).
3. If pregnancy is possible, determine the stage of gestation and estimate potential dose to embryo.
4. Substitute a nonradiologic procedure if possible.
5. If the patient is pregnant or the procedure must be done before pregnancy can be ruled out, discuss the importance of the examination and the relative risks to the mother and child with the referring physician. Record this information and a summary of your discussions on the patient's chart.
6. If the patient is pregnant and the test must be done, use the minimum amount of radiation that will supply the essential diagnostic information.

6.12 THERAPEUTIC ABORTION

Even when menstrual histories are checked before radiation is administered, occasionally it is discovered that patients were pregnant at the time they were studied. The next step is usually to question whether the pregnancy should be interrupted. That decision will depend upon:[1]

1. the radiation dose and degree of hazard to the unborn child
2. the hazard the pregnancy presents to the mother
3. the laws of the state pertaining to legal abortion

Given all these considerations, the question centers around potential hazard to the child. As stated earlier, the usual radiation dose to the embryo from diagnostic radiologic or nuclear medical procedures is one rad or less (usually much less). What is the potential for damage to the embryo from such doses of radiation? Previous portions of this chapter (see Table 6.1) have demonstrated that there is little evidence for production of major malformations after doses of less than 25 rads. Major malformations means the absence of or gross distortion of tissue or an organ or the presence of easily observable tissue hypoplasia, e.g., microcephaly or testicular atrophy. As a guide to evaluating dosage level as an indication for therapeutic abortion, the following recommendation may be helpful.

NCRP Report 54
"For any individual case, the increased risk of such effects (congenital defects) from doses below the 10-rad level received at any stage of pregnancy, according to the best knowledge available today, is very small when compared to this normal risk (4 to 6 percent of all babies are born with varying degrees of congenital defects regardless of radiation history). . . This risk (of abnormality for the unborn child) is considered to be negligible at 5 rad or less when compared to the other risks of pregnancy, and the risk of malformation is significantly increased above control levels only at doses above 15 rad. Therefore, the exposure of the fetus to radiation arising from diagnostic procedures would very rarely be cause, by itself, for terminating a pregnancy."

There appears to be *considerable* agreement that radiation doses below 10 rads have an extremely small probability of producing developmental abnormalities. As pointed out, diagnostic radiographic and nuclear medical procedures produce embryonic radiation doses which ordinarily lie well below this level, usually by a factor of 10 or more. Thus, the hazard from a *single* diagnostic procedure does not appear great enough to call for therapeutic abortion.

A possible exception to these statements would be the therapeutic use of ^{131}I beyond 10 to 12 weeks of gestation. Should a pregnancy test be negative for *early* pregnancy and the patient be treated for hyperthyroidism, the hazard to early embryo appears not to be so great as to call for abortion. Radiation dose to the early embryo depends in this circumstance upon thyroid uptake of the mother, but even in toxic patients, there would probably be less than a 5-rad fetal dose from a dose of 10 millicuries of ^{131}I administered to the mother, when the dose is delivered over several days.

The inadvertent administration of radioiodine to pregnant women beyond 10 to 12 weeks gestation (the onset of concentration of radioiodine by the fetal thyroid) has resulted in damage or destruction of the fetal thyroid. In fact, the fetal thyroid has been shown to be twice as avid for the concentration of radioiodine as the maternal thyroid gland in the second trimester of pregnancy, rising to 7.5 times as effective in the third trimester.[15] Stoffer and Hamburger[11] have presented a retrospective study of the results of inadvertent administration of radioiodine to pregnant patients for treatment of hyperthyroidism. Out of a total of 237 cases, there were six hypothyroid children from mothers treated in later pregnancy. This exceeded expectations, and based on this, the authors *strongly* recommend routine pregnancy testing prior to radioiodine therapy.

While fetal exposure during radiotherapy is uncommon (most patients are beyond childbearing years or not pregnant when treated), it occasionally happens and, in some cases, may be appreciable. If fetal presence is not contraindicated in treatment, therapeutic abortion may, unfortu-nately, be considered when fetal dose exceeds 25 rads.

SUMMARY

1. Exposure of the developing embryo to ionizing radiation may produce growth retardation, death, and/or congenital malformation.
2. The stage of development determines the organ system of totally irradiated embryos that will be most seriously damaged.
3. The most likely results of irradiation of the early embryo are death or normal development.
4. During the period of organogenesis, the possibility of induced congenital malformations is highest.
5. After the period of organogenesis, effects on the central nervous system and growth retardation are possible with higher radiation doses.
6. There is evidence that there may be increased levels of childhood leukemia and cancer following in utero exposure to diagnostic x rays.
7. The radiation dose to the embryo from diagnostic procedures in radiology or nuclear medicine is usually less than one rad.
8. Because the evidence is very supportive of the conclusion that major congenital malformations are highly unlikely to occur in the human embryo or fetus with doses less than 5 to 10 rads, inadvertent in utero exposure from a diagnostic procedure in radiology or nuclear medicine does not ordinarily call for a therapeutic abortion.

REFERENCES

1. Brent, R.L., and Gorson, R.O.: Radiation exposure in pregnancy. Curr. Probl. Radiology, *11*:1, 1972.
2. Russell, L.J., and Russell, W.L.: An analysis of

the changing radiation response of the developing mouse embryo J. Cell Physiol., *1*:103, 1954.

3. Rugh, R., and Grupp, E.: Exencephaly following x-irradiation of the pre-implantation mammalian embryo. J. Neuropathol. Exp. Neurol., *18*:468, 1959.

4. Rugh, R., Wohlfrom, M., and Varma, A.: Low-dose x-ray effects on the precleavage mammalian zygote. Radiat. Res., *37*:401, 1969.

5. Coppenger, C.J., and Brown, S.O.: Postnatal manifestations in albino rats continuously irradiated during prenatal development. Tex. Rep. Biol. Med., *23*:45, 1965.

6. Dekaban, A.S.: Abnormalities in children exposed to x radiation during various stages of gestation: Tentative timetable of radiation injury to the human fetus, Part I. J. Nucl. Med., *9*:471, 1968.

7. Kinlen, L.J., and Acheson, E.D.: Diagnostic irradiation, congenital malformation and spontaneous abortion. Br. J. Radiol., *41*:648, 1968.

8. Burrow, G., Hamilton, H., and Hrubec, Z.: Study of adolescents exposed *in utero* to the atomic bomb, Nagasaki, Japan. J.A.M.A., *192*:357, 1965.

9. Smith, E.M., and Warner, G.G.: Estimates of radiation dose to the embryo from nuclear medicine procedures. J. Nucl. Med., *17*:836, 1976.

10. Shepard, T.H.: Onset of function in the human fetal thyroid: Biochemical and radioautographic studies from organ culture. J. Clin. Endocrinol. Metab., *27*:945, 1967.

11. Stoffer, S.S., and Hamburger, J.I.: Inadvertent I-131 therapy for hypothyroidism in the first trimester of pregnancy. J. Nucl. Med., *17*:146, 1976.

12. Exss, R., and Graerve, B.: Congenital athyroidism in the newborn infant from intra-uterine radioiodine action. Biol. Neonate, 24:289, 1974.

13. Parker, H.M., and Taylor, L.S.: Basic radiation protection criteria. National Council on Radiation Protection and Measurements (NCRP) Report No. 39, 1977.

14. National Council on Radiation Protection and Measurements (NCRP) Report 54: Medical Radiation Exposure of Pregnant and Potentially Pregnant Women. Washington, D.C., 1977.

15. Book, S.A., and Goldman, M.: Thyroidal radioiodine exposure of the fetus. Health Phys., *29*:874–876, 1975.

Chapter 7

Risks of Low-Dose Irradiation

7.1 INTRODUCTION

One of the purposes of studying biologic effects of radiations is to enable medical users to consider risks incurred by irradiated persons. Whatever risks there may be can then be weighed against potential benefits, and a determination can be made of whether benefits outweigh risks. Risks of irradiation may be separated, more or less arbitrarily, into two categories—damage done by exposure to high doses and damage done by exposure to low doses. Persons exposed during diagnostic procedures (radiography and nuclear medicine) absorb small doses of radiation. Persons exposed during radiation therapy have the treated parts of their bodies exposed to high doses—other parts of their bodies receive lower doses as a result of scatter from the treatment field.

In this chapter we attempt to summarize the information presented in the previous chapters about the risks incurred by persons irradiated with low doses of radiations during diagnostic or therapeutic procedures or during the course of their work. Occupational exposure is specifically discussed in Chapter 8, and risks of exposure to high doses are discussed in Chapter 14.

7.2 BENEFITS AND RISKS

In attempting to strike a balance between risk and benefit it might seem that benefit would be an easy quantity to define. A person trained in evaluating medical benefit (a physician) must decide what he believes would happen to his patient in the absence of the use of radiation and weigh this against any possible injury which might result from the radiation. It turns out, however, that benefit is not easily defined, after all. For one thing benefit can only be defined in terms of risk (it is, in fact, the obverse of risk), and risk is not so easily defined. In most cases, benefit of irradiation is so great that, with few exceptions, the risks of exposure are heavily outweighed. This is probably true for most persons exposed during radiation therapy. If someone has a life-threatening cancer which can be controlled by irradiation, the benefit of survival seems clearly to outweigh potential hazards presented by low-dose exposure incurred by other parts of his or her body. Those hazards include increase in frequency of transmissible mutations (provided the irradiated person has childbearing expectancy) and induction of new cancers which may appear later either within or outside the treatment volume. The induction of new cancers in the treatment volume is a potential complication of radiation therapy, but its observed frequency so far is quite low. Several studies have shown that no more and probably fewer than 1% of persons treated for malignancy have grown second malignancies which can be statistically attributed to treatment. The risk of cancer induction outside the treatment volume is probably substantially less than within the volume, so the choice seems clear—control of a cancer that threatens life yields benefit greater than the risk of new cancer induction.

A difficult decision can be presented when radiation therapy must be performed on pregnant women. Radiation scattered from the treatment field can produce appreciable doses in the fetus. The potential for carcinogenesis, malformation, growth retardation (with possible associated mental retardation), and intrauterine or perinatal death is increased. Benefit versus risk here is difficult to weigh. In a sense, both mother and fetus benefit, the mother because her life may be extended and the baby for the same reason. Risk, however, seems principally incurred by the fetus. How does one weigh the *quality* of life (assuming that malformation, growth retardation, with or without mental retardation, and potential malignancies change quality of life) to be enjoyed by the baby against increased life span of the mother?

Benefit and risk of diagnostic procedures can also be difficult to assess. When the balance is between a good probability of health *restored* versus a remote probability of health *impaired*, a decision is arrived at easily. However, when the balance is between health *improved* and lives lost, choices become more difficult. For example, a fraction of malignancies induced by diagnostic radiations will be fatal. These may be among those irradiated in utero or in postnatal life. A fraction of mutations produced in gametes may produce fatal genetic diseases in those that inherit them. A fraction of chromosomal abnormalities pro-

duced in gametes may cause early death among those that inherit them. The question then becomes this: in how many people must health be improved to compensate for lives or even a single life lost? Other considerations present themselves. Did the test provide the referring physician with any more information than he or she already had? Will the results of the test *significantly* alter treatment or affect prognosis or outcome? Can benefits of tests that fail to meet these criteria balance risk of impaired health or even death?

The physician trained in the use of radiations (radiologist) should always be *convinced* that every procedure involving patient irradiation provides useful information with the least risk. It is not enough simply to follow the orders of referring physicians who frequently are misinformed or poorly informed about the information content of various diagnostic tests and their risks. The proper practice of radiology is designed to *improve* health, and that includes the health of the whole population. The proper use of radiations can accomplish this aim if each test that provides significant useful information carries a lesser risk than benefit; but if radiations are used indiscriminately, so that risk outweighs benefit in significant numbers of cases, a deterioration of the health of the population will result (Fig. 7.1).

7.3 FACTORS AFFECTING RISKS OF LOW DOSES OF RADIATION (DIAGNOSIS)

Currently in the United States, approximately one-half the population is exposed to medical radiographic procedures each year. For instance, in 1970, 76.5 million persons out of the total U.S. population of about 200 million underwent procedures involving 129 million medical diagnostic x-ray examinations.[1] There is every indication that growth in the use of medical x rays in the United States will continue at a

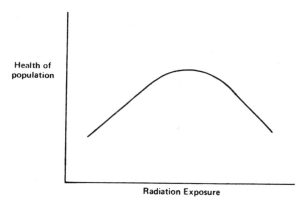

Fig. 7.1. The relationship of population health to radiation exposure (benefit vs risk). (From Payne, J.T. and Loken, M.K.[1])

rate exceeding the growth rate of the population, and that there will be an expansion in the scope and diversity of radiation techniques. Yet exposure to radiation, even in small amounts, must be considered potentially harmful.

Risks of irradiation are determined in part by the way radiation is delivered and the amount of the body irradiated. A general rule is, for any given dose, the more tissue irradiated the greater the likelihood that an effect will occur. For instance, the probability of leukemia induction will be greater when all, rather than only a part, of the bone marrow is irradiated. The rate at which radiations are delivered also affects their effectiveness. This has been mentioned in several connections so far and will be discussed more fully in Chapters 10 and 11.

Under ordinary circumstances in diagnostic medical procedures, the useful beam of x rays irradiates only *part of the body*, often a small part. The deleterious effects on the remainder of the body would be expected to be little. Dose rate is not a complicating factor either. Radiographic exposures involve a high radiation intensity for *short* periods of time. Even when multiple films are required, the exposures are within minutes of each other and cannot be considered fractioned in the sense the word is used in radiotherapy. Since dose

rate is always very rapid, no procedure is expected to be more or less injurious than another because of it.

Radiation dose is a measure of potential risk incurred by patients undergoing diagnostic x-ray studies.[2-11] Estimates of radiation dose levels for skin, bone marrow and gonads from common radiographic studies are given in Table 1.2. Skin doses for a procedure can vary from about 0.05 rad for a chest film to about 30 rads for a typical cardiac catheterization. Studies which irradiate abdomen and pelvis result in the highest gonadal doses, as would be expected.

Active bone marrow is distributed approximately as: pelvis, 36%; spine, 28%; skull, 13%; and ribs and sternum, 10%. Thus, procedures which heavily irradiate pelvis and spine, such as barium enemas, IV pyelograms, and lumbar spine examinations, result in the highest mean marrow doses. The mean marrow dose for the year 1970, averaged throughout the adult U.S. population from diagnostic x-ray procedures, was estimated to be about 0.1 rad.

7.4 RISKS OF LOW-DOSE EXPOSURE (DIAGNOSIS)

Three categories of effects are most likely to present the important risks of diagnostic exposure to radiation; carcinogenesis, mutation, and damage to the unborn. Although these have been discussed in substantial detail in previous chapters, they are summarized here for the student's convenience and for emphasis.

Genetic Damage

Potential harm to unborn generations must be a consideration in the extensive use of medical x rays on large portions of the population. Earlier apprehension about such genetic damage has lessened as knowledge has increased. The linear, no-threshold model with no repair of radiation damage has been replaced by a model which predicts reduced mutation efficiency per rad at lower doses and repair of some radiation-induced genetic damage. A number of observations support these concepts. For example, studies of descendants of mice have revealed that, even when animals were exposed to high radiation doses, offspring showed no demonstrable effects of irradiation (more than 40 generations were studied in one case). Although a single generation is not the most sensitive index of potential genetic damage, studies involving over 70,000 Japanese A-bomb survivors who became pregnant between 1948 and 1953 have shown no indication in the offspring of any gross evidence of genetic damage such as malformations, stillbirths, or neonatal deaths. With such paucity of evidence of genetic damage in man even at high doses, it is no wonder that it has not been possible to identify genetic effects at very low doses.

Since the genetically significant dose (GSD) for medical x-ray exposure is at present quite low when compared to the level from natural background (20 vs 82 millirem), this has led to the conclusion that genetic harm to the population is not the *primary* concern in the use of radiation for diagnostic purposes. This should not lead to the conclusion that mutations are harmless, only that they present a *lesser* degree of risk to the population than previously thought. However, keeping the GSD as low as possible must still be our goal for the above statement to continue to be true.

Cancer and Leukemia

The most important risk to be considered in the medical use of ionizing radiation is induction of cancer or leukemia. The risk of radiation carcinogenesis from higher doses (100 rads or more) can be approximated from several sources of epidemiologic evidence. However, there is little information on risk of tumor induction by doses of a few rads, and no reliable basis on which to infer frequency from induction at high doses.[11–13]

In the case of very low doses of radiation, it will be extremely difficult to determine by observing human populations whether or not cancer is induced. The probability of induction of cancer after doses of the order of 1 rad is, in most cases, such a small fraction of the overall cancer risk in human populations, that it is unlikely to be distinguishable from effects of other carcinogens in the environment. (The possible exception might be induction of cancer and leukemia following exposure of the embryo to diagnostic x-ray doses of 1 to 2 rads). Certainly, if cancers are being induced at these low doses, the numbers are so small that they probably add only a small risk to the *individual* patient, particularly when x-ray studies are reserved for instances in which they are a significant potential aid in diagnosis or treatment of existing disease.[15]

Intrauterine Death, Malformation and Growth Retardation

It is clearly undesirable to expose pregnant women to radiation unnecessarily, especially in the first trimester. However, there is substantial evidence suggesting that risk from radiation exposure of fetus is relatively low for diagnostic x-ray procedures. The common dose to embryo from such procedures is less than 5 rads, and usually less than 1 rad. The risk to developing embryo is described by the NCRP[14] as being "negligible at 5 rads or less when compared to the other risks of pregnancy, and the risk of malformation is significantly increased above control levels only at doses of 15 rads." Good practice indicates that one should assume that women of reproductive age are pregnant unless one knows otherwise; avoid unnecessary irradiation of the pelvis and abdomen of all women; realize that the radiation risk to the fetus is relatively low and probably never high enough to contraindicate medically needed studies; and realize that therapeutic abortion based upon radiation effects *alone* should not be considered for doses less than 10 rads.

7.5 RISKS OF RADIATION AS A SCREENING AGENT

Radiation has been used or proposed many times as an agent to detect situations perceived either as real or potential public health problems. For example, upon admission to most hospitals a chest film is routinely taken, screening, presumably, for communicable chest diseases. Repeated mammographic examinations have been suggested and even used on large asymptomatic segments of the population to screen for early breast cancer. Admittedly, the risk for any given person in a population screened this way is small, yet it is possible that such practices produce increased incidence of disease, in some cases the very diseases that they are supposed to detect. The use of radiation as a screening agent has, therefore, been questioned. Can such a practice do more harm than good? It is beyond the scope of this book to delve deeply into each individual case, and so we have chosen to present a discussion of the problems presented by mammographic screening to illustrate the general problem.

The screening of large numbers of asymptomatic women for breast cancer by mammography has been surrounded by controversy, a controversy based on the relationship of benefits to risk. Before discussing results of controlled studies of benefit of mammography in detection of early breast cancer, we will briefly summarize the risks.

The Risk of Breast Irradiation

From studies of two groups of female tuberculosis patients examined by repeated fluoroscopy, women given radiotherapy for postpartum mastitis, and Japanese women exposed to A-bomb radiations[16,17] (Table 7.1), the following general conclusions can be drawn:

1. For doses *greater* than 100 rads, there is little question that radiation induces breast cancer.
2. A linear dose-response relationship is *consistent* with the data, but because of uncertainties, other dose-response relationships cannot be excluded.
3. On the basis of a linear dose-response relationship, risk of radiation-induced breast cancer in American women is estimated to be between 6 and 9 breast cancers induced per million women irradiated per year per rad,[17] a value considerably higher than that for bone marrow irradiation and leukemia induction. For U.S. women, the lifetime natural incidence of breast cancer is about 7.58%.
4. The *minimum* latent period appears to be about 10 to 15 years with induced cases continuing to appear up to 40 years after exposure.
5. There is evidence that maximum sensitivity to radiation-induced breast cancer is found in women under 35 years of age and is reduced in older women.

Experience with Mammographic Screening

In the fall of 1963, a *controlled* study was begun to evaluate the contribution of mammography to survival from breast cancer

Table 7.1. *Risk estimates of breast cancer induced by radiation*[16]

Group	Number	Age at Mean Exposure	Mean Breast Dose in Rads	Time After Irradiation on which Risks are Based (in years)	Cancers Induced per 10^6 Women-Years per Rad
A-Bomb survivors	10,861	33	76	10–29	3.6
Nova Scotia Fluoroscopy	243	26	1215	10–30	8.4
Massachusetts Fluoroscopy	1,047	25	150	10–44	6.2
Mastitis patients	606	27	247	10–34	8.3

(this was the Health Insurance Plan of New York, or HIP study). A total of 62,000 women, 40 to 60 years of age, were divided equally between study and control groups. Although all the women in the study group were to have four annual mammographic examinations, not all completed all four. When the results were analyzed, the *entire* benefit demonstrated by the study was in women 50 years of age or older (about three-fourths of all spontaneous breast cancers occur in this age group).

In 1959, the American Cancer Society began *uncontrolled* mammographic screening programs for asymptomatic women, a program expanded in 1973 by the National Cancer Institute under the title, The Breast Cancer Detection Demonstration Project (BCDDP). Each of 27 centers was charged with examining 10,000 women annually, a total of 270,000 women each year. The protocol for this study, which enrolled women 35 years of age and older, was based upon the abrupt increase in breast cancer which occurs at about age 40, and upon improvements in mammographic techniques which might allow identification of early breast cancers in younger women than had been the case in the HIP study.

In 1976, Bailar[18] assessed the benefit/risk ratios for mammography using results of the HIP study. According to his analysis, the number of cancers found in the study group amounted to 299, of which 44 were diagnosed by positive mammographic findings alone. Based upon statistics, 14 of

the 44 patients would be expected to die of their disease during the period of the 7-year follow-up. Actually, only 2 deaths occurred. Bailar felt that the difference between 14 expected and 2 observed deaths (12) represented the best estimate of the *maximum* direct survival benefit from early detection by mammography as it was done then. In the study, 20,000 women each received an average of 3.2 sets of mammograms for an average dose of 6.4 rads (3.2 × 2 rads/examination). This represents 20,000 women × 6.4 rads or 128,000 women-rads of radiation exposure. Since the risk of radiation-induced breast cancer Bailar used was 6 cases/million women/rad/year, this figure multiplied by the 10 years following the initial 10-year latent period would predict 60 induced cancers per million women-rads. Thus, a 128,000 women-rads exposure would result in 60 × 128,000/1,000,000 or 8 cases. If the screened women lived an average of 20 years *beyond* the 10-year latent period, the total predicted number of breast cancers would be about 16. This exceeds slightly the estimated 12 deaths averted because of mammography. Bailar thus concluded that the radiation hazards appeared to be of the same order of magnitude as the benefits, and that as many cancers may be induced by radiation as are cured as a result of early diagnosis by mammography (particularly in the 35- to 50-year-old group where life expectancy would be longer than the latent period).

These conclusions were challenged on

the following bases: risks derived from linear extrapolation of effects at higher doses are questionable; reduced midline breast doses (less than 1 rad) are now possible with newer equipment; and technical advances have made possible earlier detection of tumors than was possible in the HIP study.[19,20]

The reaction to this controversy led the National Cancer Institute[21,22] to acknowledge that techniques have improved in recent years with smaller, earlier lesions at present being detected by BCDDP centers (some recent reports[20] now give *positive benefit-to-risk ratios*). Radiation doses in mammography have also been markedly decreased. Nonetheless, the Institute concluded that present data were insufficient to indicate that advances have in fact resulted in decreased mortality for women *under 50 years of age* at the time of screening.

Recommendations for Mammographic Screening

The outcome of the controversy was the decision by the NCI to place the following restrictions on use of mammography by BCDDP screening centers.[21-22]

1. Routine use of mammography in annual screening would be restricted to women 50 years of age and older.
2. Women in the age range 40 through 49 would be offered mammography screening *only* if they have had a prior history of breast cancer or if their mother or sister(s) have had breast cancer. There is little evidence that symptomatic or high-risk women are more susceptible to potential carcinogenic effects of low-level diagnostic radiation than other women.
3. For women between 35 and 39 years old, mammography would be restricted to those with a history of breast cancer.

Screening vs Diagnostic Use

The value of mammography as a *diagnostic* tool for evaluating symptoms or clinical signs of cancer (such as the presence of a lump, swelling, discharge, thickening, or other abnormality) is not questioned. Even though a dose of 1 rad might cause a slight increase in the risk of breast cancer, for individual women, the risk is small compared to the benefits. Mass screening, on the other hand, which might involve millions of women is different, since even a small risk for individuals gains significance when applied to such large numbers.

The potential hazards of mammography may appear different if the risk to an individual is considered versus the risk to large populations. For example, if a woman aged 35 received a 1-rad average breast dose from a mammographic examination, her lifetime risk of spontaneous breast cancer of about 7.58% might be expected to increase to about 7.61%. If, on the other hand, one million women received such an examination, between 234 to 425 excess breast cancers would be estimated to develop during the lifetimes of these women as a result of the radiation exposure.[17] If such large numbers are added annually, the projected numbers climb.

7.6 REDUCTION OF RISK TO IRRADIATED PATIENTS

The small risk from diagnostic x-ray studies may be reduced significantly by:

1. Limitation of areas exposed to what is essential.
2. Avoidance of undue multiplicity of views.
3. Use of appropriate filters in the x-ray machine (2 mm of aluminum will reduce skin dose by 90% in diagnostic studies).
4. Use of maximum kilovoltage compatible with the study.
5. Avoidance of fluoroscopy if the same information can be obtained from radiographs.

6. Setting up optimal technical factors and encouraging precise work to prevent technical failures which necessitate added radiation.
7. Exercising careful and conservative judgment in ordering x-ray studies.

7.7   FACTORS AFFECTING RISKS OF LOW DOSES OF RADIATION (NUCLEAR MEDICINE)

In a relatively short time, nuclear medicine has grown from a new clinical science to a major specialty. Through the use of its procedures, increasing numbers of humans are irradiated by radioactive materials localized in organs or distributed throughout their bodies. The dose of radiation they receive is a measure of the potential risk they take and must be weighed against the expected advantages of the procedure. However, because radionuclides in nuclear medicine are given internally, there are many variables, and dose and patient risk can only be estimated. Numerous factors produce uncertainties in dose calculation, particularly when applied to specific patients. Some of these, already considered in detail in Chapter 1, include elimination and redistribution of radionuclides, presence of disease states, and age of patients.

7.8   REDUCTION OF THE POTENTIAL RISK TO PATIENTS

Amount of Radionuclide

The amount of radionuclide should be kept as small as possible *consistent with obtaining desired information* from any procedure. However, nothing is worse than exposing patients to potential radiation hazards while failing to obtain proper diagnostic information because too *little* radionuclide was given. When that happens, the only alternatives are to repeat the study using more radioactivity or to ignore the data. Either course results in useless, needless radiation exposure.

Types of Radiation Emitted

Most procedures in nuclear medicine depend on detection outside the body of gamma rays emitted by radionuclides within the body. However, there are often other kinds of radiations emitted at the same time as gamma rays, which add to the dose, but because they do not contribute to the diagnosis, yield no benefit. In fact, *most* of the dose from gamma-emitting radionuclides is due, not to their gamma rays, but to beta particles, conversion electrons, and low-energy x rays emitted at the same time.

To reduce potential patient risk it is preferable to use radionuclides which emit gamma rays, lack beta particles, and have few other associated low-energy radiations. Technetium 99m (^{99m}Tc) is a good example.

Physical Half-life of the Radionuclide

Because the sensitivity of imaging equipment is limited, for some procedures there is no alternative except to administer relatively large amounts of radioactivity to assure obtaining a test suitable for diagnosis. This could result in unacceptably high radiation doses, but if the half-life of the radioactive material used is short, reasonably low dose levels can be maintained. This is true because the dose an organ receives from radionuclides deposited in it depends upon how long they are there. There is a trend toward clinical use of radionuclides with short half-lives (^{99m}Tc, half-life of 6 hours; ^{113m}In, half-life of 100 minutes), because large amounts (several millicuries), which yield good diagnostic tests while delivering relatively small radiation doses, may be given.

Prevention of Needless Irradiation

Procedures in nuclear medicine usually involve the study of concentration of radiopharmaceuticals in given organs, because the way a given compound is handled by an organ is an indication of the way the organ is functioning. Other organs may be needlessly irradiated, however, when for whatever reasons, radiopharmaceuticals are also concentrated in nontarget regions of the body, organs not involved in the study. Two means of preventing this problem are: (1) Large quantities of a nonradioactive isotope may be given to dilute the radioisotope. For instance, stable iodide (Lugol's solution) may be given *prior* to administering radiopharmaceuticals tagged with radioactive iodine. As iodine-labeled radiopharmaceuticals are degraded in the body, radioiodine is released, but the presence of large amounts of nonradioactive iodine dilutes the radioiodine and reduces thyroid uptake. (2) Agents chemically similar to particular radiopharmaceuticals sometimes decrease uptake. Examples include iodide or perchlorate to reduce concentration of ^{99m}Tc pertechnetate in body organs.

7.9 THE RISK OF DIAGNOSTIC PROCEDURES IN NUCLEAR MEDICINE

In nuclear medicine two of the most important areas receiving significant radiation doses are the *critical* organ (the organ receiving the highest dose) and the *total* body. Typical examples of dose levels from widely used diagnostic studies are listed in Table 1.3. The table shows clearly that dose levels from nuclear medicine studies are not very high. As in the case of risks from x-ray procedures, the principal potential hazards of diagnostic nuclear medical procedures are carcinogenesis, induction of leukemia, and genetic effects.

There is at present no evidence, and it would probably be very difficult to provide evidence, that any given diagnostic test using radionuclides has produced malignant tumors or genetic damage. Associated radiation doses are so low that if cancers are induced, they probably represent so small a fraction of the cancer risk normally incurred by the human population that it is unlikely the added increment could be distinguished from overall cancer incidence. In spite of the relatively high dose to thyroid (see Table 1.3), recent studies indicate no elevation in the incidence of malignant thyroid tumors in patients receiving diagnostic doses of ^{131}I.[23]

Thus, it appears that risk of cancer, leukemia, or genetic damage associated with low radiation doses delivered by radionuclides administered for diagnostic purposes is very small compared to potential benefits. Diagnostic nuclear medical procedures should be considered as having about the same potential risk as procedures in diagnostic roentgenology. This, however, is not an invitation for indiscriminate use. Risk, however small, must be weighed against potential benefits.

Women of reproductive age should be closely questioned prior to administering radionuclides to be sure they are not pregnant. Irradiation of the developing embryo may occur from radionuclides concentrated in maternal organs or in the embryo itself if the radionuclide crosses the placental barrier.

Because the number of diagnostic nuclear medicine procedures is small compared to the number of diagnostic x-ray studies performed, and because there is a relatively low gonadal dose from these examinations (see Table 1.3), the contribution of nuclear medicine to the GSD from medical procedures is a very small part of the total. Therefore, nuclear medicine diagnostic studies probably do not represent a significant risk of genetic damage to the population.

The risks of diagnostic procedures in nuclear medicine, however small, must be related to benefits of such procedures to determine whether they are worthwhile. This is not always easy to do. Adelstein[24] has attempted such an assessment for lung scans to determine if the number of patients saved outweighs the risk of radiogenic leukemia. Using a linear extrapolation from cases of leukemia induced by high radiation doses, he predicted that a patient undergoing a lung scan (0.1 rad to bone marrow) would have a risk of induced leukemia of about 3 in 1,000,000. Assuming that mortality for untreated pulmonary embolism and for treated pulmonary embolism is 20% versus 5% respectively, if the scan is responsible for the initiation of treatment in only one out of 50 patients examined for pulmonary embolism, a salvage rate of 3 per 1000 is achieved. This is a thousand times *greater* than the projected risk of leukemia, giving a benefit/risk ratio of 100:1. But, is a lung scan *essential* to diagnosis? Although it is only one of a number of tests for pulmonary embolism, in the absence of lung scans, it is likely that at least 1 out of 50 patients seen in emergency rooms with symptoms of pulmonary embolism would be sent home on the basis of normal radiographic and other findings. The lung scan is a good discriminator since a normal scan essentially rules out pulmonary embolism, while an abnormal scan is a firm basis for hospital admission. Obviously, benefit versus risk relationships are more difficult to evaluate where indications for and utility of a procedure are less clear-cut.

7.10 HAZARDS OF TREATMENT OF HYPERTHYROIDISM WITH [131]I

Use of ^{131}I for treatment of hyperthyroidism was introduced in 1942 and has been widely accepted as a successful means of controlling the disease since about 1950. Patients treated number in the hundreds of thousands. Most problems concerning possible hazards center around the potential for the induction of thyroiditis and hypothyroidism, thyroid and other abnormalities in the unborn, and leukemia and thyroid carcinoma.

Thyroiditis and Hypothyroidism

Radiation thyroiditis is an acute condition occurring within two weeks after exposure to radiation and is characterized by inflammation and eventual necrosis of some or all thyroid cells. *Clinically significant* radiation thyroiditis[25] is highly unlikely at radiation doses below 20,000 rem from [131]I.

Hypothyroidism occurs with Graves' disease not just following ^{131}I therapy but even after surgery and has a cumulative incidence of about 20% in 10 years. Long-term incidence seems to depend on dose of ^{131}I with the cumulative incidence over 10 years varying between 30% and 70%, depending on dose.[25]

Pregnancy and [131]I

Since iodine crosses the placenta, treatment of thyrotoxicosis with ^{131}I in pregnant patients beyond 10 weeks gestation has resulted in hypothyroid children. Use of pregnancy tests in potentially pregnant patients who are to receive therapeutic levels of ^{131}I is recommended. Surveys have indicated that inadvertent ^{131}I therapy for hyperthyroidism during the first trimester did not result in increased stillbirths or malformations as compared to the incidence in a similar number of random pregnancies. Hypothyroidism did result however, if pregnancy had gone beyond the first trimester.

Genetic Effects and [131]I

The radiation dose to ovaries or testes is less than 3 rads from treatment of hyperthyroid patients with a dose of 10 mCi of [131]I. Since this is a dose within the range given by many common diagnostic roentgenologic procedures, it seems as unreasonable to withhold ^{131}I treatment from young men and nonpregnant young women, on grounds of genetic hazard alone, as it would be to withhold diagnostic x-ray examinations.

Induction of Thyroid Cancer

Many epidemiologic studies have identified the thyroid gland as particularly susceptible to radiation-induced cancer. Most of these studies, however, have associated x rays rather than beta radiations from ^{131}I with increased cancer incidence. Some of the reasons for this apparent lack of carcinogenicity may include lack of uniformity of dose with ^{131}I, dose-rate differences, and higher doses associated with [131]I.

In view of the large number of patients treated with ^{131}I, the reported numbers of thyroid cancers certainly do not indicate much of an excess above chance expectations, even allowing for a long latent period. Most studies fail to show any excess, but a recent review of the subject[25] reports a very small increased *absolute* risk of thyroid cancer from ^{131}I in children compared to adults (0.06 cases per million per year per rem as opposed to 0.05 cancers per million per year per rem in adults). External irradiation of thyroid is 70 times more effective at increasing absolute risk. Children were, however, two to three times more susceptible than adults to ^{131}I-induced *benign* nodules.

Induction of Leukemia

Another possible side effect of ^{131}I treatment for thyrotoxicosis is induction of leukemia. Sporadic case reports in the 1950s of leukemia occurring in patients treated with radioiodine led to two large-scale studies. In 1960, Pochin[26] analyzed data from several countries involving 59,200 patients. His estimate was that 21 cases would occur by chance in such a sample, and this was higher than the 18 cases reported to have occurred.

In 1968, Saenger et al.[27] reported the results of a cooperative study involving 18,400 patients treated with ^{131}I as compared to 10,700 treated with surgery for thyrotoxicosis. When the leukemia incidence in radioiodine-treated patients was compared to surgically treated patients, no difference was demonstrated, although the incidence of leukemia in *both* groups was greater than in the population at large. At least this much seems clear, there is no evidence for the induction of large numbers of leukemias from that particular kind of radioiodine therapy.

Benefit vs Risk

In selecting any form of treatment, one always must try to assess available alternatives, risks incurred, and benefits expected. Surgical treatment, even in experienced hands, results in permanent hypoparathyroidism or vocal cord damage in about 1% of patients. Radioiodine treatment provides rapid, effective, inexpensive, and permanent control of Graves' disease without the risks of surgery. The disadvantages have been enumerated and include a progressively increasing incidence of hypothyroidism, the possibility (probably quite small) of increasing incidence of thyroid cancer, and the possible production of genetic damage.

7.11 TREATMENT OF THYROID CANCER WITH [131]I

Because of the high doses of ^{131}I used in treatment of thyroid cancer, rather sizable radiation doses in various body parts result (a 100 mCi dose of ^{131}I results, for example, in a red marrow dose of about 30 rads). Ultimately, bone marrow is depressed, and this depression is the limiting factor in treatment. During therapy, complete blood counts, especially emphasizing lymphocytes, should be routine.

Pochin[26] has documented increased risk of cancer associated with high-dose therapy. In some 200 patients treated with radioiodine, he observed that 12 malignancies occurred when 5.2 were expected; that 4 of these were acute leukemias while the expected number of acute leukemias was less than 1; and that there were 4 cases of carcinoma of the breast when only 1 was expected. All these differences were highly significant. For all other malignancies, there was no difference between observed and expected number. When one considers the risk of carcinogenesis just noted and the expected benefits from treatment of what can be a serious, life-threatening disease, the risks appear justified. In spite of large radiation doses given to treat thyroid cancer, other studies have indicated no evidence of genetic damage, even in patients given several hundred millicuries of ^{131}I.[28]

7.12   TREATMENT OF POLYCYTHEMIA VERA WITH ^{32}P

It has been known since 1945 that acute leukemia occurs in patients treated with ^{32}P for polycythemia vera. However, acute leukemia also occurs in patients with polycythemia vera treated without radiation, using only drugs or phlebotomy. (A dose of 4 mCi of ^{32}P delivers a dose of about 100 rads to bone marrow, liver and spleen; the remainder of the body receives about 10 rads.) It is uncertain whether the observed leukemia incidence following ^{32}P therapy is the result of therapy or of other factors. It is possible that ^{32}P therapy, by prolonging life, permits the natural evolution of polycythemia vera into acute leukemia or that the possible carcinogenic action of ^{32}P results from irradiation of a susceptible population. A controlled study by the Polycythemia Vera Study Group is now under way to provide more definitive answers to this question.

SUMMARY

1. Diagnostic x irradiation is limited to the portion of the body which is to be viewed. Efforts such as limitation of field size, and gonadal shielding reduce radiation exposure to parts of the body outside the beam.

2. The GSD from medical x-ray exposure is about 20 mrem/year.

3. The greatest potential risk from medical x-ray exposure is believed to be induction of cancer.

4. The risk of irradiating embryos with doses of less than 10 rads appears to be negligible when compared to the other risks of pregnancy.

5. The value (benefit) of diagnostic radiography for evaluating symptoms or clinical signs of disease is not questioned. In given individuals it almost always exceeds the putative attendant risk.

6. When x rays are used for mass screening, even the small risk carried by individuals gains significance when applied to large numbers. Under such circumstances, unless benefits are quite significant, they may be exceeded by potential risks.

7. Radiation doses received by individual patients from nuclear medicine procedures can only be estimated because of such variables as patients' size, age, and pathology, as well as redistribution of the radiopharmaceutical.

8. The contribution of nuclear medicine procedures to the GSD is small compared to the contribution from x-ray procedures.

9. There is no evidence that any given diagnostic test using radionuclides or x rays has produced malignant tumors. Benefits of most such procedures seem to far outweigh the small putative risk usually predicted.

10. Risk of induction of thyroid cancer after treatment of hyperthyroidism appears to be very low.

11. It does not appear that the low ^{131}I doses associated with the treatment of hyperthyroidism have induced detectable numbers of cases of leukemia, but it does appear that the higher doses of ^{131}I used to treat thyroid cancer have induced leukemia.

Table 7.2. *Estimates of low-dose risks*

Genetic Defects

First Generation—1 rem of parental irradiation.

Number affected per million liveborn*	(5–75) 30	BEIR, 1980 UNSCEAR, 1977

At equilibrium—1 rem of parental radiation over many generations.

Number affected per million liveborn*	(60–1,100) 185	BEIR, 1980 UNSCEAR, 1977

*Spontaneous incidence—107,000 per million liveborn

Cancers Induced

Additional cancer deaths in the lifetime of 1 million people after low doses of radiation, per rad*	(77–226) 100	BEIR, 1980 UNSCEAR, 1977

*Spontaneous—160,000 lifetime fatal cancers per million

Developmental Abnormalities

Embryonic dose less than 5 rads*	negligible
Embryonic dose greater than 15 rads*	significant increase of malformation

*Spontaneous incidence—4 to 6% of births.

12. It is still unclear whether ^{32}P therapy for polycythemia vera induces leukemia.

13. Table 7.2 summarizes *estimates* of risks of low-dose irradiation.

REFERENCES

1. Payne, J.T., and Loken, M.K.: A survey of the benefits and risks of the practice of radiology. CRC Crit. Rev. Clin. Radiol. Nucl. Med., Cleveland, CRC Press, July, 1975.
2. Gonad Doses and Genetically Significant Dose from Diagnostic Radiology U.S., 1964 and 1970. HEW Publication FDA 75-8034, U.S. Government Printing Office, Washington, D.C., April, 1976.
3. Rosenstein, M.: Organ Doses in Diagnostic Radiology. HEW Publication FDA 76-8030, U.S. Government Printing Office, Washington, D.C., May, 1976.
4. Trout, E.D., Kelly, J.P., and Cathey, G.A.: The use of filters to control radiation exposure to the patient in diagnostic roentgenology. Am. J. Roentgen., 67:946, 1952.
5. Rueter, E.G.: Physician and patient exposure during cardiac catheterization. Circulation, 58:134–139, 1978.
6. Gitlin, J.N., and Lawrence, P.S.: Population Exposure to X-Rays U.S. U.S. Public Health Service Publication No. 1519. U.S. Government Printing Office, Washington, D.C., 1964.
7. Antoku, S., and Russell, W.J.: Dose to the active bone marrow, gonads, and skin from roentgenography and fluoroscopy. Radiology, 101:669–678, 1971.
8. Rogers, R.T.: Radiation dose to the skin in diagnostic radiography. Br. J. Radiol., 42:511–518, 1969.
9. Shleien, B., Tucker, T.T., and Johnson, D.W.: The mean active bone marrow dose to the adult population of the United States from diagnostic radiology. Health Phys., 34:587–601, 1978.
10. Gregg, E.C.: Radiation risks with diagnostic x rays. Radiology, 123:447–453, 1977.
11. National Research Council. Advisory Committee on the Biological Effects of Ionizing Radiations (BEIR III): The Effects on Populations of Exposure to Low Levels of Ionizing Radiation. National Academy of Sciences, Washington, D.C., 1980.
12. Brown, J.M.: Linearity vs non-linearity of dose response for radiation carcinogenesis. Health Phys., 31:231–245, 1976.

13. Cohen, B.L.: The cancer risk for low-level radiation. Health Phys., *39*:659–678, 1980.
14. National Council on Radiation Protection and Measurements (NCRP) Report No. 54: Medical Radiation Exposure of Pregnant and Potentially Pregnant Women. Washington, D.C., 1977.
15. Oosterkamp, I.W.J.: Benefit/risk comparisons in diagnostic radiology. Medicamundi, *21*:1–6, 1976.
16. Boice, J.D., et al.: Risk of breast cancer following low-dose radiation exposure. Radiology, *131*:589–597, 1979.
17. National Council on Radiation Protection and Measurements. (NCRP) Report No. 66: Mammography. Washington, D.C., 1980.
18. Bailar, J.C.: Mammography: a contrary view. Ann. Intern. Med., *84*:77–84, 1976.
19. Swartz, H.M., and Reichling, B.A.: The risks of mammograms. JAMA, *237*:955–965, 1977.
20. Fox, S.H., et al.: Benefit/risk analysis of aggressive mammographic screening. Radiology, *238*:359–365, 1978.
21. National Institutes of Health/National Cancer Institute: Consensus development meeting on breast cancer screening: Issues and recommendations. J. Natl. Cancer Inst., *60*:1519–1520, 1978.
22. Statement on recommendations of the Consensus Development Panel on Breast Cancer Screening. J. Natl. Cancer Inst., *60*:1523–1524, 1978.
23. Hohn, L.E., Lundell, G., and Wallinder, G.: Incidence of malignant thyroid tumors in humans after exposure to diagnostic doses of iodine-131. Retrospective study. J. Natl. Cancer Inst., *64*:1055–1059, 1980.
24. Adelstein, S.J.: Radiation risks. *In* Nuclear Medicine. Edited by H.N. Wagner, Jr. New York, HP Publishing Co., 1975.
25. Maxon, H.R., et al.: Ionizing irradiation and the induction of clinically significant disease in the human thyroid gland. Am. J. Med., *63*:967–978, 1977.
26. Pochin, E.E.: Long term hazards of radioiodine treatment of thyroid cancer. *In* Thyroid Cancer. UICC Monograph series, Vol. 12, Berlin, Springer-Verlag, 1969, pp. 293–304.
27. Saenger, E.L., Thomas, G.E., and Tompkins, E.A.: Incidence of leukemia following treatment for hyperthyroidism. Preliminary report of the Cooperative Thyrotoxicosis Therapy Follow-up Study. JAMA, *205*:855–862, 1968.
28. Sarker, S.D., Beierwaltes, W.H., Satinder, P.G., and Cowley, B.J.: Subsequent fertility and birth histories of children and adolescents treated with I-131 for thyroid cancer. J. Nucl. Med., *17*:460–464, 1976.

Chapter 8

Risks of Occupational Exposure

8.1 INTRODUCTION

People whose work requires exposure to ionizing radiation rightly wonder what risks they take. They are concerned about possible harm to themselves, their children, and their potential offspring. Their question is, what potential harm is associated with *chronic* low-dose exposure? No exact answer is known to this question— no one can say with precision how much risk is run by any given occupationally exposed individual.

The current "standard" or guide for the occupationally exposed is the maximum permissible dose (MPD) or the radiation protection guide. This gives specific upper levels of permitted occupational exposure. Adherence to the guide carries, presumably, good assurance that whatever risk there may be will be very small.

8.2 THE MEANING OF THE MPD

The detrimental effects of radiation in man may be classified as either stochastic or nonstochastic.[1] *Stochastic* effects are those for which the probability of an effect occurring, rather than the severity of the effect, is a function of dose without a thresh-

old. Examples include genetic effects and carcinogenesis. *Nonstochastic* effects are those for which the severity of the effect varies with dose and for which there is a *threshold* (erythemas, epilation, cataracts, cell depletion of the bone marrow and impaired fertility are examples). The aim of radiation protection should be to *prevent* detrimental nonstochastic effects and to *limit* the probability of occurrence of stochastic effects to levels deemed acceptable. The MPD levels are, in fact, based upon the concept of an "acceptable risk" and are derived from a long experience of exposure of human beings to levels of ionizing radiation that have not demonstrated deleterious effects. As the International Commission on Radiation Protection (ICRP) has said:[2]

"The permissible dose for an individual is that dose, accumulated over a long period of time or resulting from a single exposure, which, in the light of present knowledge, carries with it a negligible probability of severe somatic or genetic injuries; furthermore, it is such a dose that any effects that ensue are more frequently limited to those of a minor nature that would not be considered unacceptable by the exposed individual and by competent medical authorities.

"Any severe somatic injuries (such as

leukemia) that might result from exposure of individuals to the permissible dose would be limited to an exceedingly small fraction of the exposed group; effects such as shortening of life span, which might be expected to occur more frequently, would be very slight and would likely be hidden by normal biological variation. The permissible dose can therefore be expected to produce effects that could be detectable only by statistical methods applied to large groups."

The fact that when dose levels remain within protection guides no effects are observed does not necessarily mean that none occurs. Epidemiologic studies as currently performed may not measure effects of low radiation doses or distinguish them from the background of spontaneously occurring diseases in the population. With improved techniques or better samples, some damage may be demonstrated in the future, but the absence of observed effects in existing studies makes it unlikely that future studies will yield positive results.

In the past, MPD recommendations were limited to doses that were small fractions of the doses at which adverse effects had been observed. For example, if leukemia could be demonstrated to occur in a statistically significant percentage of a population following bone marrow doses of low-LET radiations of 100 rads, the *permissible* dose for occupationally exposed persons would be set at a very small fraction of 100 rads so that the risk per person in the population would be extremely small.

The ICRP has recently used a different approach, namely "acceptable risk," to assess levels of risk associated with the MPD. The Commission judged acceptability of the level of presumed risk of radiation work by comparing it with "other occupations having high standards of safety," those in which the average annual mortality does not exceed 1 death per 10,000 individuals. As it happens this approach results in the same MPD as that arrived at by the older approach (5 rem/year).

Another useful measure of the hazard of occupational exposure is decrease in workers' life expectancy compared to workers in other occupations. Table 8.1 makes such a comparison. It is well to recall that most medical personnel receive far less than the MPD (the BEIR III estimate is an average of 300 to 350 mrem per year per worker). Obviously the risks are less than for a number of other occupations, including several regarded as low-risk jobs.

8.3 THE EVOLUTION OF LIMITS FOR OCCUPATIONAL EXPOSURE

The evolution of the MPD is summarized in Table 8.2. Prior to 1928 there was no accepted unit for expressing radiation dose, but by the early 1930s the roentgen (R), a unit of exposure to x rays and gamma rays, was established. In 1934 an MPD of 0.2 R/day was recommended by the ICRP, whereas the National Committee on Radiation Protection (the NCRP) in the United States recommended a level of 0.1 R/day to correct for dose measurements made in air instead of on the skin.

Because of knowledge gained and the advent of new types of radiations, the ICRP reduced the level from 0.1 R/day to 0.3 R/week in 1950. However, at about that time the genetic aspects of radiation exposure came under intensive study. Because of these studies, in 1956 and 1957, the ICRP and NCRP (for the purpose of holding down genetic dose to a maximum of 60 rem to the age of 30 years among the increasing numbers of radiation workers) recommended a further reduction. The result is the present MPD level[5] of 5 rem/year (0.1 rem/week) which applies to workers above age 18. Specific body parts (such as hands and forearms) have higher MPD levels [4,5] than do more radiosensitive and vital body organs. Radiation exposures resulting from necessary medical and dental procedures are *not* included in the determination of the radiation status of occupationally exposed

Table 8.1. *Approximate days of life expectancy lost as a result of hazards from occupational radiation exposure compared with hazards in other industries, based upon exposure for remainder of working life[3]*

Occupation	Age (at beginning of exposure) in years.			
	20	30	40	50
Deep-sea fishing	1393	923	551	273
Coal mining	155	103	61	30
Coal & petroleum products	111	74	44	22
Railway employment	96	63	38	19
Construction	94	62	37	18
All manufacturing	21	14	8	4
Paper, printing, publishing	12	8	5	2
Radiation work at 5 rem/year	68	32	12	3
Radiation work at 500 mrem/year*	7	3	1	0.3

*BEIR III estimate of the average dose for medical personnel is 300–350 mrem/year.[4]

Table 8.2. *Chronology of the maximum permissible dosage*

Group and Dates	MPD Level Rem/Year
Prior to 1934	100
ICRP (1934–1950)	60
NCRP (1935–1948)	30
ICRP (1950–1956)	15
ICRP (1956–)	5

persons and are *not* included in an individual's occupational exposure history.

Doses of 25 rem and below can generally be expected to produce no symptoms, detectable clinical findings, or impairment of efficiency. Therefore, an accidental or emergency dose of up to 25 rem occurring only once in a lifetime can be neglected in the determination of occupational radiation status.

The need to minimize exposure of embryos is the controlling factor in *occupational* exposure of pregnant women and limits the MPD of the fetus during the entire gestational period to 0.5 rem.

The dose limits for the *general* population from radiation have been set at one-tenth of that of the occupationally exposed, since much larger numbers of people are involved, including children and pregnant women.

8.4 THE ALARA CONCEPT

For many years the operating philosophy in radiation protection was that occupational radiation exposures should be kept as low as practicable, but the vagueness of this concept permitted wide interpretation. This led the National Radiation Council (NRC) to the "as low as reasonably achievable" (ALARA) philosophy. Under this approach, each exposure situation should be evaluated not in terms of how one might avoid exceeding MPD values but in terms of how to achieve the *lowest* exposure commensurate with reasonable cost and effort. This does not mean disagreement with the concept that occupationally exposed workers may safely be exposed to MPD levels of radiation. However, the ALARA concept requires these limits be used as a *ceiling* and *not* as desired operating conditions.

8.5 THE RISK OF CANCER AND LEUKEMIA FROM OCCUPATIONAL EXPOSURE

The carcinogenic effect of x rays in irradiated animals and man was the first late somatic effect of ionizing radiation observed, and is the *only significant potential somatic injury at levels within the MPD*. Most data on induction of cancer by radiation in man have involved extremely high, localized doses, the accumulated dose being in the order of hundreds of rads. Occupational exposure, in addition to being at a low dose level, is protracted over a long time period, a factor which may substantially reduce the risk. The ICRP[1] believes that the calculated rate at which fatal malignancies might be induced by occupational exposure to radiation up to the MPD "should not in any case exceed the occupational fatality rate of industries recognized as having high standards of safety."

8.6 THE RISK OF LIFE SHORTENING

Most studies in animals have shown shortened life expectancy following radiation exposure. It is uncertain whether the effect is an acceleration of the aging process or the result of induction of neoplasms. Recent studies of human and animal populations, however, suggest that life shortening is the result of radiogenic malignancies. Examination of the condition of superficial blood vessels of A-bomb survivors as an index of the aging process[6] showed no significant difference between control and exposed groups. By this criterion, radiation exposure did not accelerate aging.

The dose levels received by present-day radiation workers are unlikely to be causing significant shortening of the average life expectancy (see Table 8.1). Radiation-induced life shortening in man has been reported in a study of the average life expectancy of radiologists.[7] The cause of death and length of life among radiologists were compared to the general population, physicians in general, and other medical specialists. The study concluded that in the *past* the life expectancy of radiologists was significantly shorter than that of other physicians or that of other United States white males (Fig. 8.1). Over the years, the *average*

life span of American radiologists has increased as the MPD has decreased. At this time, and since 1960, no life shortening of radiologists has been detected. Shields Warren[7] concluded that, "the present maximum permissible dose level therefore appears well established as safe for protection against somatic effects."

8.7 THE RISK OF GENETIC DAMAGE

As stated previously, a major reason for the reduction in the MPD to 5 rem per year in 1956 to 1957 was concern about genetic damage. Many experts now agree that the present MPD levels represent a *very small* genetic risk, but it should be remembered that any radiation exposure is *potentially* damaging from a genetic standpoint.

8.8 THE INDUCTION OF CATARACTS

Exposure of the optic lens to ionizing radiation causes cataracts in man and animals. This effect has a clear threshold at about 200 rads for a single exposure and 500 rads for exposure protracted over several weeks (for low-LET radiation). When doses are spread over extended time periods as in normal occupational exposures, available data suggest that for x rays, the dose required to induce cataracts is in the range of several hundred rads. On this basis, the induction of vision-impairing cataracts would not appear to be possible as long as the radiation levels remain within the present MPD value of 5 rem/year.

Neutrons are particularly efficient for cataract production. This became apparent in 1949 when 10 nuclear physicists with a common history of exposure to cyclotron radiation were shown to have incipient cataracts. By 1950 there were 21 cases of cataracts among cyclotron workers. This increased effectiveness for cataract formation by fast neutrons is compensated for by using considerably higher (a factor of 10)

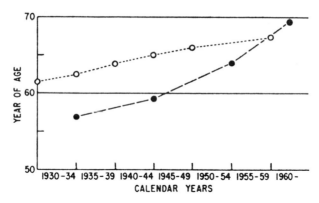

Fig. 8.1. Mean age at death by calendar year for radiologists (●—●) and for U.S. white males (○—○) over 25 years of age. (From Warren, S.[7])

QF values for fast neutrons in the determination of radiation dose levels given in rems.

8.9 THE POSSIBILITY OF IMPAIRED FERTILITY OR STERILITY

There is no indication that radiation doses within the MPD values have any effect on sexual capacity (libido or potency) or that they cause sterility.

SUMMARY

1. Radiation protection guides or maximum permissible dose (MPD) values have been derived on the basis of experience. The levels of risk associated with the MPD appear to be in keeping with those of industries generally recognized to have high safety standards.
2. The present MPD values were derived from the concept that there is *no* threshold for cancer induction, and a concern about genetic damage to the growing numbers of occupationally exposed individuals.
3. In spite of failure to observe effects at low doses, it is assumed that there is some degree of damage to human beings, no matter what the exposure level.
4. The MPD is a guide and does *not* represent a line which is safe on one side and not on the other. It represents, rather, an "acceptable risk," a risk that is assumed to be present but is difficult to demonstrate by scientific studies.
5. All radiation exposure should be minimized as much as is practical.

REFERENCES

1. International Commission on Radiological Protection (ICRP) Publication 26: Recommendations of the International Commission on Radiological Protection. New York, Pergamon Press, 1977.
2. Taylor, L.S.: Radiation protection standards. Radiology, 74:824, 1960.
3. Reissland, J., and Harries, V.: A scale for measuring risks. New Scientist, 83:809–811, 1979.
4. National Research Council, Committee on the Biological Effects of Ionizing Radiations (BEIR III): The Effects on Populations of Exposure to Low Levels of Ionizing Radiation. National Academy of Sciences, Washington, D.C., 1980.
5. National Council on Radiation Protection and Measurements (NCRP) Report No. 43: Review of the Current State of Radiation Protection Philosophy, Washington D.C., 1975.
6. Tsuya, A., et al.: Capillary microscopic observation on the superficial minute vessels of atomic bomb survivors. Hiroshima, 1972–73. Radiat. Res., 72:353–363, 1977.
7. Warren, S.: The basis for the limit of whole-body exposure: experience of radiologists. Health Phys., 12:737–742, 1966.

Chapter 9

Cell-Survival Curves

9.1 INTRODUCTION

To understand cellular radiation injury, especially as it applies to radiation therapy, it is important to study cell survival as a function of radiation dose and to study the factors that modify dose-survival relationships.

9.2 DOSE-SURVIVAL RELATIONSHIPS

When populations of nearly any mammalian cell type are irradiated in single exposures with a number of varying doses of low-LET radiations, a graph of the results produces sigmoid survival curves. If the same thing is done with very high LET radiations, exponential curves are observed (Fig. 9.1).

In sigmoid dose-response relationships, slope is not constant. Through a low-dose range, the slope of survival as a function of increasing dose is shallow, but through higher dose ranges, the slope is steep. There is a point where the shallow slope ends and the curve inflects (bends in), beginning the steeper slope. This inflection is generally called a "shoulder" (see Fig. 9.1).

Dose-survival relationships produced by very high LET radiations are not inflected. The slopes of these curves are constant (see Fig. 9.1). Careful examination of such curves (Fig. 9.2) reveals that for *any given dose increment*, the *same proportion* (not number) of an irradiated cell population is killed. In any properly performed irradiation, all cells in the irradiated population are exposed to the same dose of radiation. The same *number* of ionizations and excitations must, on the average, occur in each of them. Yet, as can be seen in Figure 9.2, given doses kill a fixed proportion of the irradiated population, and a fixed proportion always survives. The reason for this is supposed to be that when each cell in an irradiated population is exposed to the same amount of radiation energy, an injury resulting in death is produced in only a proportion of them. In other cells, the *particular* injury that causes death is not produced. Lethal radiation damage appears to be the result of a *single*, critical, radiation-caused event called a *hit*. There can be no question of accumulation of damage; a *single* hit causes death. Dose-response curves have a constant slope (no inflections), and cell killing appears to be a simple, exponential dose function (see Fig. 9.2).

In contrast to exponential survival curves obtained with high LET radiations, sigmoid curves produced by low LET radiations have two slopes; they are inflected. The slope of the initial region of the curve is less steep than that past the inflection, indicating a *lesser* efficiency for cell killing in the dose range beneath the shoul-

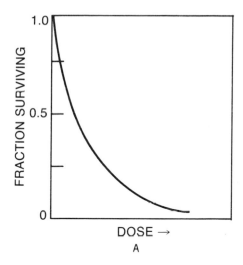

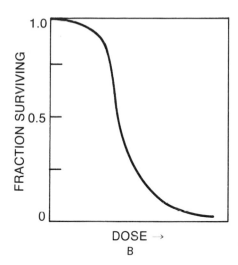

Fig. 9.1. Hypothetical examples of *typical* dose-survival relationships. *A.* An exponential relationship. Surviving fraction of cells varies as an exponent of dose. Cell survival is a constant exponent of dose. *B.* A sigmoid relationship. Initially the curve is relatively flat, but with a negative slope. There is an inflection (called a shoulder)—a dose range in which the curve changes slope and turns downward. The shoulder is followed by the long exponential terminal segment in which survival is a constant exponent of dose.

der than above it (Fig. 9.3). In the example shown in Fig. 9.3, it can be seen that 90% of irradiated cells survive 200 rads of radiation, and 10% are killed. However, if 200 more rads are given to the survivors, 38% are killed and a smaller proportion of the population survives. While the chance of surviving the first 200 rads is 9 in 10, the chance of surviving 200 *additional* rads is only about 1 in 2. Clearly, the first 200 rads killed some cells but only injured others, increasing the risk of killing them when they were exposed to 200 rads more. The efficiency of the radiations for cell killing is *less* in the low-dose range (0 to 200 rads kill only 1 in 10) than it is in higher ranges (200 to 400 rads kill nearly half the surviving irradiated population).

Damage that injures but does not kill is called *sublethal.* It may accumulate and reach lethal levels. The inflection, the shoulder of sigmoid survival curves, is a kind of threshold. It is the point at which (or the dose at which) most surviving cells in the population have accumulated maximum sublethal damage.

Slopes of sigmoid survival curves in dose ranges *above* their shoulders can be experimentally determined with reasonable precision. When irradiated cell populations are homogeneous, the slope appears to be constant and is an exponential dose function. For various reasons, however, the slopes of survival curves in the dose range *beneath* the shoulder are much more difficult to determine. Yet knowledge of dose responses of cells in that dose range is, from the point of view of medicine, the more important to obtain. During medical procedures, the bodies or parts of bodies of human beings are rarely irradiated (in single exposures) with doses greatly in excess of 200 rads. With photon irradiation, the shoulders of many mammalian cell-survival curves occur typically at about 200 rads. Lethal cellular responses to irradiation following doses of *less* than 200 rads are of the greatest relevance to medicine yet are least known. This deficit is probably of special significance in radiotherapy. The aim of radiotherapy is to kill or sterilize as many cells of malignant growths as possible while sparing those of surrounding normal tissues. The efficiency for killing cells of var-

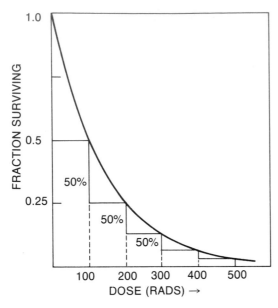

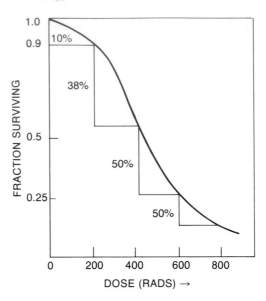

Fig. 9.2. A hypothetical dose-survival curve obtained using very high LET radiations. In this example 50% of the irradiated population survives 100 rads. If 100 rads *more* are given the survivors, 50% again survive. If 100 additional rads are given to those surviving, 50% survive. That relationship, 50% survivors after each 100 rads increment, continues throughout the full extent of the dose range. *In general*, the same *proportion* (in this example, 50%) of an irradiated cell population survives or is killed by any given dose increment (in this example, 100 rads).

Fig. 9.3. A hypothetical sigmoid dose-response relationship. In this example 90% of irradiated cells survive 200 rads; 10% are killed. If those survivors are given 200 rads more (cumulative dose 400 rads), 62% survive, 38% are killed. If 200 rads more are given those survivors (cumulative 600 rads), 50% survive. If 200 rads more are given those survivors, 50% survive. Thereafter, 50% will survive 200 rads, no matter how much radiation has already been given.

ious kinds of tissue (malignant and normal) by radiation doses in the dose range below the shoulder is crucial to successful treatment. If survival curve slopes were to vary substantially among various cancers and normal tissues, it would mean that the efficiency of *given* doses would differ for killing cells of different tissues. Depending on *how* efficiency for cell killing varied, more or fewer cancer cells than normal cells might die after exposure to particular doses of particular radiations, and this would set the probability of successful cancer control.

While the exact shape of survival curves in the region below the shoulder is disputed, many, perhaps most, investigators believe it to be exponential or nearly so. It seems that sigmoid survival curves may be described as follows. In the dose range *below* the shoulder, survival varies approximately as a simple exponential dose func-

tion of shallow slope; in the dose range *above* the shoulder, survival varies as an exponential dose function of steeper slope. If this is the case, the two mechanisms described in the following paragraphs possibly account for mammalian cell killing by ionizing radiations.[1-4]

In the dose range less than the shoulder, the nearly exponential nature of these curves suggests that cell death occurs as a result of a *single* lethal event, one hit. It appears that a small fraction of any kind of mammalian cells, even after exposure to the lowest doses, is killed as a result of a single hit. No threshold seems to exist for cell killing.

The fact that efficiency for cell killing increases after a certain dose (see Fig. 9.3), however, suggests that through low-dose ranges, the single lethal hit does not occur in many cells. Nevertheless, damage takes place. This *sublethal* damage accumulates and reaches lethal levels (see Fig. 9.3).

Thus, in the dose range above the shoulder, the curve is also exponential, cell death in this dose range being due also to a single critical event, the production of a *final* injury making the level of damage lethal.

No known explanation exists for the dual radiation-dose response, but there are several possibilities. It is possible that within mammalian cell populations, there are two kinds of cells, one having only a single sensitive region or structure that results in death if damaged and another having several regions or structures all of which must be damaged to result in death. This explanation, while not disproven, is not commonly accepted. An alternative hypothesis seems much more likely. It is supposed that, with few exceptions, all mammalian cells have several sites or structures called targets. To cause cell death all of these targets must, at the same time, be injured by irradiation in some critical or specific way. When cells die after a low radiation dose in the range less than the shoulder, it is because for some reason, *all* their *targets were hit simultaneously by one* or very few radiations. When cells die after higher doses in the range above the shoulder, it is because *all* their *targets were simultaneously hit by several* of the large number of radiations passing through them.

The explanation supporting this latter hypothesis is as follows. As low-LET radiations pass through matter, their energy is dissipated unevenly along their tracks. Energy dissipation is concentrated near the end of the track. The rest of the track, which is its greatest part, consists of clusters of ions each at a substantial distance from each other. Cells lying along most of the track are liable to incur damage from only a few ionizations *from any given radiation* because ion density is low. The chances are that only a fraction of their targets will be hit by any given radiation. Therefore, given radiations are likely to produce sublethal damage but not cell death. However, cells lying at or near the *end* of these tracks are subject to dense ionization. There is a

strong possibility that *all* the targets they possess will be hit at once and that such cells will die. Death is due to a single critical event, the simultaneous inactivation or injury of all the targets by the dense ionization at the end of the low-LET radiation track.

Cells lying along the less dense part of radiation tracks may also be killed if all their targets are simultaneously hit, but a reasonable chance of that occurring exists only when radiation dose is sufficiently high that dense ionization is caused by a large number of radiations.

When radiation dose is low, cell killing is mainly the result of dense ionization at the ends of tracks. Since the ion-dense portion of low-LET tracks is quite short, few cells are killed and efficiency for cell killing is low. Remaining cells may be sublethally injured because only one (or a few) of their targets is hit by sparse ionization along the principal part of the track.

The effect of increasing dose is to increase the number of cells lethally injured because more radiations mean more tracks and more ion-dense tails. It also means greater degrees of sublethal injury will have been produced in survivors since more radiations mean greater numbers of ionizations per cell and more targets hit per cell. A dose must eventually be reached which is sufficiently large that, in most surviving cells, all targets save one are simultaneously hit. This would represent maximum degrees of sublethal damage for that cell type, and a single additional hit ensures that all targets will have been simultaneously destroyed, an event that is lethal. On survival curves, the point of maximal sublethal damage is the shoulder. The slope change observed after the shoulder, i.e., the increased efficiency of given doses for cell killing in the dose range above the shoulder, occurs because *both* mechanisms now are killing cells. In some cells all targets are simultaneously hit by the ion-dense track tails of the numerous radiations (dose is high), and in others, all targets are simultaneously hit by the less dense portions of tracks of several radiations passing through them at once.

The production of *hits* is an exponential function of dose, so cell survival, which may depend on the production either of one or a number of hits, may either be exponential (if a single hit causes death) or sigmoid (if several hits must accumulate to cause death). The number of targets that must be simultaneously inactivated for cell death to occur is a parameter determining whether cells die from or survive irradiation. The number of hits or critical events needed to inactivate those targets is symbolized by the letter n. That value is not constant; it varies according to various conditions. Conditions which favor the production of many simultaneous target inactivations, such as high-LET irradiation, result in low values for n, and those favoring the production of few simultaneous target inactivations, like low-LET irradiation, result in higher values for n. Consequently, as LET of radiations used to irradiate cells is increased, n should decrease, because increased ion density along the track increases the probability of several simultaneous target inactivations. This is, in fact, commonly observed. For given multitargeted cell types, the highest values for n are noted when the lowest LET radiations are used, and the lowest value of n is found when very high LET radiations are used. It must also be true that, as LET of radiations is increased, lesser degrees of sublethal damage are done by given doses, because if these radiations traverse cell nuclei, they inactivate several targets at once. If very high LET radiations are used, the ion density along their tracks is so high that if they pass through cell nuclei, they inactivate all the targets simultaneously. No sublethal damage occurs, and cells die as a result of a single critical event, the passage of a very high LET radiation. The survival curves they yield are exponentials of a single slope and have no shoulders.

Values of n, the number of hits needed

to inactivate all targets simultaneously, are derived from survival curves. This notion is most easily illustrated by plotting cell-survival curves on semilogarithmic coordinates (Fig. 9.4). On such coordinates exponential relationships appear as straight lines of a *single* slope, but sigmoid relationships are still shouldered, the shoulder being an inflection between low and high efficiency for cell killing.

As has already been stated, the exponential *beyond* the shoulder describes the relationships between increasing dose and the production of the *last* in a series of a number of hits. It is assumed that the relationship between increasing dose and the production of the other hits in that series is the same. Therefore, the slope of hit production versus increasing dose for *all* hits in the series will be the same. The expo-

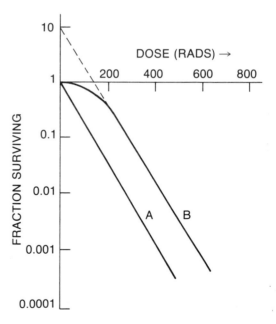

Fig. 9.4. Hypothetical exponential and sigmoid dose-survival relationships plotted on semilogarithmic coordinates. The abscissa (x axis) is a linear coordinate, but the ordinate (y axis) is a logarithmic coordinate. On such a coordinate system, exponential relationships appear as straight lines. The fully exponential curve (curve A) intercepts the y axis at 1.0. Consequently n = 1. The exponential segment of the sigmoid relationship (curve B) must be extrapolated to the y intercept. In this example, the intercept is 10 and n = 10.

nential *beyond* the shoulder may, therefore, by extrapolated backward to zero dose, the point where it intercepts the y axis. That point will be n, the *extrapolation number,* the *average* number of hits needed to inactivate simultaneously all targets in that cell population. Where low-LET radiations and mammalian cells are concerned, n is usually greater than one, because mammalian cells are, on the whole, multitargeted, and low-LET radiations are not effective at producing many simultaneous target inactivations.

9.3 RADIOSENSITIVITY

The extrapolation number is one parameter controlling cell survival. Another is *radiosensitivity,* the amount of radiation energy (dose) required to make a hit. For reasons that are not entirely clear, different cell types vary in radiosensitivity. Under identical conditions and using the same radiations, mammalian cells having the *same* extrapolation numbers may require different doses of these radiations to produce a given level of effect. In a hypothetical example, two populations of mammalian cells having values of n equal to 10 may require different x-ray doses to kill 90% of them. The population that is reduced to a 10% survival level by the lesser dose is, for some reason, more sensitive to x rays than the other. Since both populations have extrapolation numbers of 10, both require the same number of hits to inactivate all their targets simultaneously. It is clear that, because more radiation is required in one instance to inactivate all targets simultaneously, more radiation must be needed to produce each hit.

Radiosensitivity is given by the *slope* of survival curves and is the *dose of radiation which produces,* on the average, *one hit per cell in an irradiated population.* It is designated D_o, and it is the *37% dose slope of the exponential segment* of survival curves beyond the shoulder (Fig. 9.5). D_o can be de-

termined only from the exponential segment of survival curves beyond the shoulder, because it has to do with hit production. It is helpful to remember that radiosensitivity (D_o) refers to the dose required to produce hits and not necessarily to the dose needed to kill cells. The dose required to kill cells will be determined by the *two* parameters n and D_o—the average number of hits that must be produced as well as the dose of radiation needed to produce (on the average) a hit.

It may already be clear from the foregoing that if D_o is given to a cell population, not every cell in the population will have sustained one hit. Radiation-matter interactions are random. When D_o is given to a cell population, some cells will have had one hit, some will have been hit several times, and some not at all. The result is that, when D_o is administered, not every cell dies. In fact 37% survive, because not

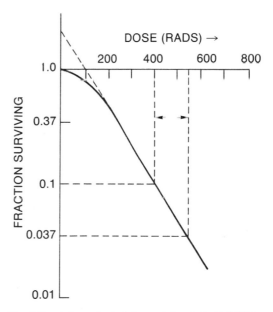

Fig. 9.5. A hypothetical sigmoid dose-survival curve. The dose of radiation which reduces the *number of survivors* to 37% (the 37%-dose slope) is D_o. It is the dose which, on the average, produces one hit per cell in the irradiated population. Because *hits* are produced as exponential functions of dose, D_o can be taken only from the exponential segment of the curve beyond the shoulder. In the example used in the figure, D_o is 150 rads.

enough hits are made in them to kill them, and of the 63% which do not survive, some cells are hit more times than is necessary to kill them.

The exponential nature of dose-survival relationships has important implications for radiotherapists. It means that no radiation dose, *irrespective of how large*, guarantees that every cell in an irradiated population will be killed. If successful control or cure of human cancers depends on eradicating the entire malignant population, radiotherapy will not be able to assure this. Presumably, no dose can be large enough to ensure accomplishing this end, and furthermore, the size of the dose that may be administered is limited by normal tissue tolerance. On the other hand, successful control of human cancers by reducing the malignant cell population to a small enough number that the cancer does not regrow sufficiently to be a medical problem in the patient's remaining lifetime might be a realizable goal.

9.4   THE QUASI-THRESHOLD (D_q)

The inflections or shoulders of survival curves have already been referred to as a kind of threshold. They are the point at which sigmoid curves acquire their final slope and the point at which cells of an irradiated population have absorbed maximum sublethal damage. They are given the designation D_q and are called *quasi*-thresholds. D_q is the *dose* of radiation needed to bring an irradiated population to the shoulder of its survival curve and a measure of the dose of radiation which, in most cells, produces mainly sublethal damage (Fig. 9.6).

Doses below D_q kill few cells per unit of dose and above it kill more cells for the same unit of dose. It must be remembered, however, that D_q is a *quasi*-threshold. Doses below D_q do kill cells, and in fact, there does not seem to be a threshold for cell killing. Efficiency per unit of dose for cell

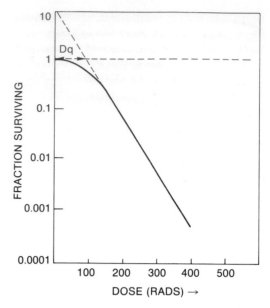

Fig. 9.6   A hypothetical survival curve demonstrating the *quasi*-threshold (D_q). The point of intercept between the exponential and the survival axis of unity (100%) is the quasi-threshold. It is a measure of the width of the shoulders of survival curves. Doses of radiation below D_q produce, *on the whole*, sublethal damage; doses above produce, *on the whole*, lethal damage.

killing may vary, but even the lowest doses kill some cells.

Neither n, D_o, nor D_q is constant. They can and do vary according to cell type, physiologic state of cells, type of radiation, dose rate of radiation, and the presence or absence of radioprotectors or sensitizers.

9.5   INTERPRETATION OF SURVIVAL DATA

At present it is not clear what happens in cells when they are irradiated that produces the typical survival curves. Two major hypotheses have been advanced and each has its proponents.

Target Theory

This theory, which has already been described in some detail, proposes that, in cells, there are sites or regions called targets that control various cellular functions.

When radiations make critical changes in these targets, they may make them inactive. Critical changes can be produced either directly (radiation interacts with the target) or indirectly (free radicals or other radiation-produced products interact with the target). If only one target controls a function, a single critical change in that target will inactivate it, and loss of function will result. If *several* targets control the same function, then each of them must sustain a critical change and be inactivated for loss of function to result. The amount of radiation energy needed to inactivate a target may vary from cell type to cell type indicating that the *sensitivity* of targets to radiation injury varies.

In mammalian cells, sublethal damage appears possible, which suggests that one critical event does not *lethally* damage most of them. If the target theory is correct, there must be several targets *within* mammalian cells, but what these targets might be is unknown. Many investigators think DNA is the leading candidate. If one or a few changes in DNA are made, the effect is likely to be sublethal. This is not to say that nothing at all has occurred. Damage to DNA may certainly result in mutations or other far-reaching effects, but this damage need not be *lethal*. If a critical level of damage or changes in DNA is produced, it may be lethal. The region of the survival curve *before* the shoulder may represent doses of radiation that produce increasing levels of sublethal DNA damage in most cells. The shoulder may be the dose region where maximal amounts of sublethal DNA damage has taken place, and the part of the curve beyond the shoulder may represent the dose region where the response of cells to only one more change in DNA will be lethal.

Other workers believe the chromosomes to be the principal targets for radiation. This larger genetic milieu seems to them a better candidate for the putative targets on biologic grounds, but the effects on cells would be the same as outlined above for DNA.

Q Theory

A second hypothesis which has been proposed to explain survival data can be called Q theory. Q theory does not postulate a series of accumulated hits to explain the shapes of survival curves. Instead this hypothesis suggests that most mammalian cells are equipped with an as yet unidentified substance called "Q" which either protects them from radiation damage or repairs radiation injuries before they can be expressed. In a low-dose range (below the shoulder, D_q) there is plenty of this substance. Consequently, few cells are killed by irradiation in that range. As dose is increased, more and more radiation interactions occur in cells, and the supply of Q decreases, because, in the process of protection or repair, it is used up. A dose will be reached at which so many interactions have occurred in irradiated cells that all Q will be gone. The *next* interaction can then produce damage which is lethal. The dose at which all Q is exhausted is D_q, the shoulder of survival curves.

The Q hypothesis, then, seems to suggest that radiation interactions in cells may produce damage that is potentially lethal, but that the presence and action of Q either prevent such damage or convert such lesions to nonlethal lesions. The hypothesis does not suggest that the action of Q completely reverses radiation damage in every instance. It simply proposes that Q may prevent lesions that could be lethal from occurring, or it may convert potentially lethal lesions to ones that are nonlethal. Thus, it does not obviate the possibility of radiation-produced nonlethal cellular lesions such as mutation or carcinogenesis.

Of these hypotheses, target theory is most commonly invoked to explain radiation-survival curves. It must be stressed, however, that so far, unequivocal evidence of the correctness of any theory has not been produced.

SUMMARY

1. Cell survival is described by two parameters, n and D_o.
2. D_o is the radiation dose which, on the average, produces one hit per cell in an irradiated population. It is determined from the *slope* of exponential segments of survival curves and is the dose that reduces the number of possible hits to 37%.
3. The average number of hits per cell in a population is called n (the extrapolation number). It is the intercept between exponential survival curves and the Y axis.
4. D_q is the *quasi*-threshold and is the dose of radiation which measures the width of survival-curve shoulders and the maximum amount of sublethal damage possible in an irradiated population.
5. Neither n, D_o nor D_q is a fixed unit. All may vary according to a variety of factors—some intrinsic to cells and some dependent upon radiation characteristics.
6. An alternate hypothesis to target theory is the Q hypothesis which also explains survival data.

REFERENCES

1. Neary, G.J.: The dependence of the oxygen effect on the intensity of gamma irradiation on *Vicia Fabia*. *In* Progress in Radiobiology. Edited by J.S. Mitchell, B.E. Holmes, and C.L. Smith. Springfield, Illinois, Charles C Thomas, 1955, p. 355.
2. Alper, T., editor: Cell Survival after Low Doses of Radiation: Theoretical and Clinical Implications. New York, John Wiley and Sons, 1975.
3. Elkind, M.M.: Fractionated dose radiotherapy and its relationship to survival curve shape. Cancer Treat. Rev., 3:1–15, 1976.
4. Elkind, M.M.: The initial part of the survival curve. Implications for low-dose, low-dose-rate radiation responses. Radiat. Res., 71:9–23, 1977.

Chapter 10

Modifiers of Cellular Response to Irradiation

10.1 INTRODUCTION

The relationship between radiation dose and cellular damage is different under different conditions. Various factors such as LET of radiations, degree of oxygenation, and dose rate during irradiation, significantly change the amount of cellular damage done by given radiation doses.

10.2 LET (LINEAR ENERGY TRANSFER)

There is an important relationship between LET of radiations and production of cellular radiation damage. In brief, if radiations having various LETs are used to irradiate populations of identical cells under identical conditions, given doses of these radiations will not produce the same effects or degree of effect. The general relationship between LET and cellular response is that as LET of radiations is *increased*, the efficiency or effectiveness of those radiations for producing biologic effects is also *increased* until maximum effectiveness is reached. Radiations with LETs *less* than those that produce an effect at the maximum frequency are relatively *less* effective or efficient than those that produce an effect at maximum frequency. Also, radiations with LETs *greater* than those that produce an effect at maximum frequency are relatively inefficient because they spend *more* energy than is needed to produce an effect at maximum frequency (see Fig. 2.2).

10.3 SURVIVAL DATA

Mammalian cell-survival data reflect the general relationship discussed in the previous section (Fig. 10.1). The inflection on curves produced by very low LET radiations (x and gamma rays) occurs only after substantial radiation doses have been absorbed by the cell population. This indicates that in the dose range *before* the inflection, whatever damage is done kills few cells. It can be understood as a failure of low-LET radiations in that dose range to inactivate sufficient targets in most cells to kill them. It may also be that low-LET radiations do so little damage in that dose range that sufficient quantities of Q are present in most cells either to repair these lesions or to convert them to nonlethal lesions.

Very high LET radiations, on the other hand, produce survival curves without inflections (see Fig. 10.1). This indicates that, if a radiation passes through the nucleus of a cell, it kills that cell. Depending on

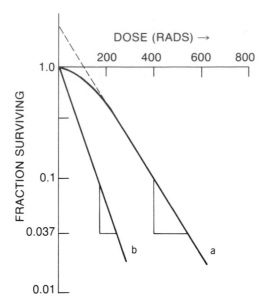

Fig. 10.1. A hypothetic example of the results of irradiation of identical mammalian cells with varying doses of very low LET radiations (a) and very high LET radiations (b). Very low LET radiations produce survival curves with substantial shoulders, indicating significant degrees of sublethal damage are produced. Very high LET radiations produce curves without shoulders, indicating only lethal damage is produced. The slopes of survival curves also differ, D_o for low-LET radiations is large relative to that for high-LET radiations.

shoulders of survival curves decreases (see Fig. 10.1). Since D_o is the radiation dose which, *on the average*, produces one hit per cell in an irradiated population, it indicates that high-LET radiations are more effective at producing hits than are low-LET radiations. The modifying influence of LET can be summarized as follows. The shoulders of survival curves become narrower and their slopes steeper as a function of increases in LET. In turn, this means that the proportion of sublethal to lethal damage done by *given* doses of radiation will be dependent on LET. It will be greater with low- than with high-LET radiations.

interpretation, it would indicate that any very high LET radiation, passing through a cell nucleus, has a high probability of hitting and simultaneously inactivating *all* the targets or of using up all the Q. High-LET radiations, therefore, are relatively effective at cell killing because, depending on point of view, they produce many simultaneous hits or many potentially lethal lesions which, because of repair or conversion, instantly exhaust the supply of Q.

As LET of radiations increases, the parameter n decreases. Very low LET radiations normally yield survival curves with large values of n; very high LET radiations usually yield survival curves with n equal to one; and radiations of LETs between extremes often yield values of n greater than one but less than those observed with very low LET radiations. In addition, as LET increases, D_o of the exponential beyond the

10.4   THE OXYGEN EFFECT

The response of mammalian cells to irradiation may be different, depending on whether they are at normal degrees of oxygenation or hypoxic during irradiation. Cells which are fully oxygenated are more sensitive to radiation than hypoxic cells. Mammalian cells are supposed to be fully oxygenated when they are at an oxygen tension of approximately 20 to 40 mm Hg. Cells at oxygen tensions less than 20 mm Hg may be defined as hypoxic, but significant changes in radiosensitivity begin to be observed only when cells are extremely hypoxic, with oxygen tensions between 0 and 3 mm Hg.

10.5   DOSE-MODIFYING EFFECT OF OXYGEN

The effect of oxygen has usually been described as dose modifying, that is, the shape of survival curves using well-aerated and hypoxic cells is the same, but the slopes of the curves differ (Fig. 10.2). The extrapolation number (n) remains unchanged, but the dose of radiation needed to produce, on the average, one hit per cell (D_o) is greater in hypoxic than in normally aerated cells.

The dose-modifying or enhancing action of oxygen is expressed by the oxygen enhancement ratio (OER). The radiation dose required to produce a given end point in given cells under hypoxia is compared to the dose of the same radiations required to produce the same end point in the same kind of cells when they are well aerated.

$$OER = \frac{\text{Radiation dose producing end point in hypoxia}}{\text{Radiation dose producing end point in normal aeration}}$$

Since oxygen is a sensitizer, the latter dose is usually smaller than the former, and the ratio is usually greater than one. The actual

value of the OER, however, is dependent on LET. For very low LET radiations (x and gamma rays) the OER is between 2.5 and 3.0, i.e., two and one-half to three times the dose of low-LET radiations will be needed to produce various end points in hypoxic cells as is needed to produce the same end points in normally oxygenated ones. If radiations of higher LET are used, the OER will be lower; for very high LET radiations, the OER will be very much lower.

When oxygen acts merely as a dose-modifying agent, the D_o of the survival curve is the only characteristic of the curve that changes, and the OER does not vary with the size of the dose. The OER at very high doses is the same as it is at lower doses.

Not all experimenters studying oxygen's effect agree that D_o is the only parameter altered. A reduction in extrapolation number (n) as well as D_o has been reported[1-9] in cells under extreme hypoxia or in hypoxia for protracted periods. In such cases the presence or absence of oxygen is not simply dose modifying; it also affects the relative amounts of sublethal injury produced (Fig. 10.3). In cases where extrapolation number is reduced, the OER will not be constant for all dose levels but will vary with the size of the dose.

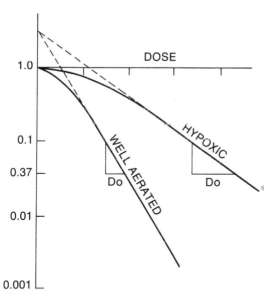

Fig. 10.2. An illustration of the dose-modifying effect of oxygen. The aerated and hypoxic curves have the same extrapolation number, but D_o, the slope, is clearly greater in the hypoxic than in the well-aerated populations. Oxygen has made the cells more sensitive, and in this example, D_o of the hypoxic population is three times that of the well-aerated one. Curves such as these are typical of oxygen enhancement by low-LET radiations. The difference in doses required to reduce the surviving population to various survival levels (0.1, 0.01, 0.001, etc.) is a factor of 3, a factor commonly observed using low-LET radiations against well-aerated and hypoxic mammalian cells of the same type.

10.6 OER AND LET

The size of the oxygen enhancement ratio depends on LET. Generally, high-LET radiations produce smaller values of OER than do low-LET radiations, and the OER declines as LET of radiations increases (Fig. 10.4). This means that the presence of oxygen greatly sensitizes cells to low-LET radiations (OER is between 2.5 and 3.0), but as higher LET radiations are used, it makes less and less difference whether cells are well aerated or hypoxic; the difference in response becomes more uniform. When very high LET radiations such as alpha par-

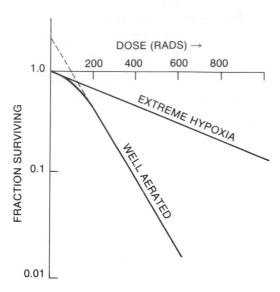

Fig. 10.3. A hypothetic example of reduction in extrapolation number (n) in extremely hypoxic cells. In the well-aerated population, n is significantly greater than 1.0, but in extremely hypoxic cells, the survival curve is exponential, and n = 1.0. The "shouldered" nature of the well-aerated curve suggests sublethal damage is produced, but in extreme hypoxia, this does not happen.

ticles are used, the OER equals 1.0; that is, there is no difference in response between hypoxic and oxygenated cells.

10.7   THE ACTION OF OXYGEN

It is not clear how oxygen enhances radiations' effects. One explanation suggests that the presence of oxygen makes *less likely* the restoration of damaged organic molecules after irradiation. An example follows:

1) $RH \xrightarrow{\text{radiant energy}} R\cdot + H\cdot$
2) $RH \longleftarrow R\cdot + H\cdot$ restoration
3) $R\cdot + H\cdot + O_2 \longrightarrow RO_2\cdot + HO_2\cdot$
no restoration

A key organic molecule (RH) absorbs radiation energy. The free radicals $R\cdot$ and $H\cdot$ are formed. Because free radicals "seek" electrons with opposing spins to satisfy electron spin imbalance, and because the free radicals $R\cdot$ and $H\cdot$ have electrons with opposite spin *and* are close together, there is a high probability that they will react to reform RH. RH is restored, and this critical molecule is not lost. In the presence of oxygen, $R\cdot$ and $H\cdot$ have a high probability of interacting with oxygen to form the free radicals $RO_2\cdot$ and $HO_2\cdot$. RH cannot be restored. Moreover, $RO_2\cdot$ and $HO_2\cdot$ may react further, potentially to attack other organic molecules. Under hypoxia, $R\cdot$ and $H\cdot$ have a higher probability of restoring RH than under normal aeration, and less radiation damage per unit of dose (fewer hits/unit of dose) will be done.

The process just described doubtless occurs whether high- or low-LET radiations are used, but the actions of high-LET radiations are less susceptible to enhancement than those of low-LET radiations. High-LET radiations are, in themselves, very effective at producing damage and *very* high LET radiations are so effective that the passage through a cell by one of them is lethal.

10.8   CLINICAL SIGNIFICANCE OF OXYGEN EFFECT

The oxygen effect has no known practical significance in diagnostic radiology or diagnostic nuclear medicine. Radiations used are low LET, but doses are so low that little cell killing occurs. The chief significance is in radiotherapy. It seems likely that most, if not all, cancers contain cells which are hypoxic and some which probably are severely hypoxic. All animal cancers tested have hypoxic components, and strong indirect evidence of hypoxic cells in human cancers has been produced by analysis of cancer histology[10] and by calculation.[11] In addition, more direct evidence of hypoxic cells in human cancers was obtained during a test of an hypoxic-cell sensitizer in patients.[12]

Hypoxic cells in cancers are presumed to

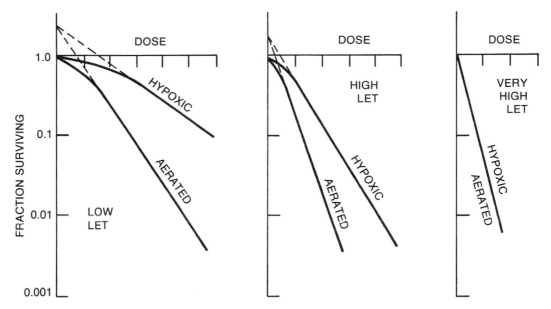

Fig. 10.4. A hypothetic example showing decrease in the OER with increasing LET. At left, using very low LET radiations (photons), the OER is about 3.0. In the center, high-LET radiations (such as 14 MeV neutrons) produce an OER of about 1.7. At right, using very high LET radiations, such as alpha particles, hypoxic and aerated curves are superimposed. The OER is 1.0.

come about as a result of the poorly organized manner of malignant growth. For some reason, during periods of active growth, cancers produce more cells than they lose. A mass forms in which the malignant tissue grows and accumulates faster than the vascular system needed to support it. In time, cancer cells find themselves at a great distance from capillaries. Oxygen and nutrient diffusing from these capillaries is used up by nearby cells and not enough for the needs of distant cells remains. Some become hypoxic and malnourished (Fig. 10.5).

The presence of hypoxic, malnourished cells in cancers may well be a problem of considerable dimensions for radiotherapists. Part of the problem is obvious. Since hypoxic cells are radioresistant (especially to low-LET radiations such as x and gamma rays), and since cancers are likely to have hypoxic cells in them, there will be more difficulty in killing or sterilizing these cells than cells of normal tissue. If sufficient photon radiation were used to kill or sterilize hypoxic cancer cells, the radiation tolerance of surrounding normal tissues would, in many instances, be exceeded. This results in reactions and complications of such severity that they present considerable problems in clinical management. If radiation doses sufficient to kill or sterilize only well-aerated tumor cells are given, normal tissue tolerance is not exceeded, but the possibility of recurrence in the irradiated field (failure to control locally) is raised. Reference to Fig. 10.5 helps explain why this happens. Many cells nearest capillaries that are well aerated (proliferating) may be killed by irradiation, but many of those more distant (hypoxic and quiescent or severely hypoxic and quiescent) may survive. After cells in the well-aerated, proliferating zone are killed, fewer will be present to take up oxygen and nutrient emanating from the capillary. Presumably, because there are fewer cells to compete, oxygen and nutrient diffuse farther, and some cells in the quiescent hypoxic zone become aerated and better nourished. If they receive enough oxygen and nutrient, they may resume proliferation. There is an irony

DEAD CELLS—ANOXIA

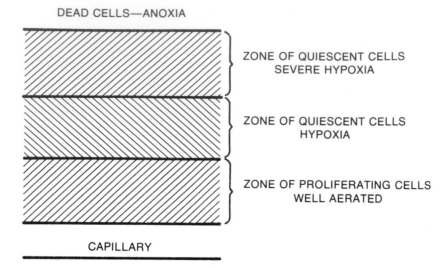

ZONE OF QUIESCENT CELLS
SEVERE HYPOXIA

ZONE OF QUIESCENT CELLS
HYPOXIA

ZONE OF PROLIFERATING CELLS
WELL AERATED

CAPILLARY

Fig. 10.5. A schematic representation of regions of cancers near capillaries. Nearest the capillary, cells are normally aerated and proliferative. Newly formed cells push others away from the capillary. At a distance, too little oxygen and nutrient, which must diffuse and which are consumed by cells over which they diffuse, remain to support proliferation. This results in a zone of hypoxic, probably malnourished, quiescent cells. Continued production of new cells near the capillary pushes older cells even farther away. Hypoxia and malnutrition become progressively more severe. A zone of severely hypoxic, malnourished cells is formed. Finally, cells too far from feeding capillaries have no oxygen, become anoxic, and die.

here. In *untreated* cancers, hypoxic and severely hypoxic cells probably do not threaten life since they cannot proliferate and they ultimately die. Life is threatened by proliferating cancer cells, because they cause continued cancer growth. In *treated* cancers, proliferating cells are killed and do not threaten life, but hypoxic cells may re-oxygenate, begin proliferating, and become potentially life threatening. Clearly, if local control of cancers with radiation is to succeed, eradication of the hypoxic component is of the highest importance. In the absence of eradication, hypoxic cells may be the focus of recurrence and the cause of therapy failure.

10.9 FRACTIONATION

Various means have evolved or have been developed to cope with hypoxic cancer cells. Fractionation of radiation dose during therapy is one. When dose is fractioned, the first few fractions probably kill mostly well-aerated, proliferating cells.

This makes available more oxygen and nutrient for the hypoxic component, and in theory at least, these should re-oxygenate. It is uncertain whether hypoxic human cancer cells actually do re-oxygenate, but experiments performed in animals have shown that re-oxygenation of hypoxic cancer cells does occur during fractionated radiotherapy.[13] As the regimen of fractionated radiotherapy progresses, more and more hypoxic cells re-oxygenate (some at each fraction) and acquiring oxygen, become well aerated and radiosensitive. Ideally, a course of radiotherapy consisting of a sufficient number of fractions, spaced so there is enough time between each one to permit re-oxygenation, would lead to radiosensitization of all hypoxic cells in a tumor.

Whether or not human cancer cells re-oxygenate is not known, but the possibility is supported by the fact that local control so often results after radiotherapy. It is believed that nearly all common cancers have hypoxic foci. If cells in these foci retain hypoxia-related radioresistance throughout courses of radiotherapy, recurrence *in the treated field* should be a common experience. In fact, the contrary is true.

10.10 HYPERBARIC OXYGENATION

An approach relying on delivering greater quantities of oxygen to all regions of cancers by increasing oxygen tensions in capillaries has been given the shorthand name, "hyperbaric oxygenation." Oxygen concentrations in capillaries depend on two factors—the concentration of oxygen in inspired gas, and the pressure of inspired gas.

By having patients breathe pure oxygen (oxygen is only 21% of air) and by increasing the pressure of inspired oxygen to three atmospheres, concentrations of oxygen in capillaries theoretically become approximately 2000 times greater than when patients breathe air at atmospheric pressure. Using oxygen at high pressure (hyperbaric oxygen) then ought to increase capillary oxygen and this, in turn, should make more oxygen available to *all* cells in given cancers. Little effect is anticipated on well-aerated cells, cancer or normal, because radiosensitivity does not change appreciably as cellular oxygen tensions increase above 20 mm Hg, but cells that are hypoxic may re-oxygenate and become sensitized. In persons breathing hyperbaric oxygen, given doses of low-LET radiations ought to kill or sterilize more cancer cells than in those breathing air at atmospheric pressure. Some experiments carried out in patients have shown that the technique in fact resulted in fewer recurrences in treated fields.[14] However, other experimenters trying the method[15] observed no such advantage. In any case, even those experimenters who observed improved local control did not observe a significant improvement in overall survival. Of course, *survival* is not necessarily related to local control, except where disease has not spread. The equivocal results of these experiments and the failure to affect overall survival resulted in loss of interest in the method. Recently however, interest has quickened because a probable explanation for the differences in the results of the experiments has emerged. Those claiming success for the method generally did not anesthetize their patients while those experiencing failure did. It is now believed that the anesthetics were radioprotectors and may have made all cells in the cancers resistant to radiation, even those that re-oxygenated under hyperbaric oxygenation.

Also, in the intervening years, the ability to control disseminated disease has improved, so that better control of primary disease is a desirable goal, even in cases where disease has spread.

10.11 HIGH-LET RADIATIONS

The interrelationship between high-LET radiations and the OER makes exploitation of high-LET radiations to eradicate hypoxic cells in cancers an obvious goal. Since the OER is smaller with high-LET radiations than with low-LET radiations, given doses of high-LET radiations ought to kill greater numbers of hypoxic cancer cells than the same doses of low-LET radiations. In theory, very high LET radiations would be best because they have an OER of 1.0 (see Fig. 10.4), but *penetrating* beams of such radiations are very difficult to obtain. Radiations that have an OER significantly smaller than the 3.0 of x and gamma rays and are, nevertheless, available in penetrating beams, have been sought. So far, neutron beams are the choice most often made. Depending on their energy, neutrons have OERs in the range of 1.5 to 1.8 (about half that of x or gamma rays), and again depending on their energy, have penetrating qualities comparable to cobalt 60 therapy sources and linear accelerators. In addition to the lower OER, they should have other therapeutic advantages, among them a smaller shoulder on the survival curve. Because their higher LET reduces both shoulder and OER, given doses of neutrons should reduce cell numbers in an irradiated cancer substantially more than the same doses of x or gamma rays. A smaller proportion of hypoxic cells would be resistant to them (OER = 1.7; OER of x or gamma rays = 3.0), and more cells would be lethally injured using given doses of neutrons compared to the same doses of x or gamma rays. Rapid reduction in cell number should lead to rapid re-oxygenation and sensitization of the remaining tumor mass. Fractioned neutron therapy would be expected to produce a greater overall shrinkage of cancers and better control of irradiated lesions than similar regimens of low-LET radiations.

A number of clinical trials using neutrons have been conducted over the years. The results to date can be summarized as follows. Neutrons seem to offer significant advantage, more for some tumor types than others, but a great deal of work remains to be done. Experimentation using various fraction patterns, doses, dose rates, and combinations of neutrons and low-LET radiations must be carried out. Nevertheless, promise is there and may well be fulfilled in years to come.

While it is true that high-LET radiations have a lower OER than low-LET radiations, it is not automatically true that this advantage can be easily exploited clinically. Charged particles are a case in point. LET is dependent on mass and charge. The more massive and/or highly charged a particle is, the greater will be its LET. In turn, the greater the LET, the smaller will be the OER. The absorption pattern of charged particles as they pass through matter shows first a track of relatively low LET, then a region of track in which LET steeply increases (the particle is rapidly losing energy), then a peak of extremely high LET (the Bragg peak), and then no further energy absorption (the particle is at rest). Figure 2.1 in Chapter 2 describes this. The region of highest LET (the Bragg peak) is the portion of particle track in which the OER would be smallest and the particle most effective at killing or injuring hypoxic cells. Unfortunately, Bragg peaks are generally very narrow, encompassing only a few millimeters. Thus, very high LET radiations are not generally useful clinically, because the most desirable region of track (the peak) falls far short of the space occupied by even the smallest detectable cancers. Various devices (principally filters) can be used to spread peaks out. In so doing, they are broadened, but they are also lowered and flattened. LET is correspondingly lowered and the OER goes up. For example, a radiation producing an LET in its peak that gives an OER of nearly 1.0 might give an OER of about 2.0 in a peak that is spread out. A *clinically useful peak* (a spread-out peak) may have an OER not as good as neutrons though somewhat better than x or gamma rays. While the difference between an OER of about 3.0 (x and gamma rays) and 2.0 (spread-out peaks) is significant, the enormous cost of producing clinically useful beams of very high LET radiations may not be justified by this degree of improvement in the OER.

At present, interest in charged radiations lies in the *dose distribution* attainable with them. In the matter of beam shaping and dose distribution they are superior to photons. Radiation fields can be very precisely defined, and because of that, normal tissues are more successfully spared while higher doses are given to tumor tissue. They offer special promise in treating cancers situated in normal tissues whose damage or destruction is particularly dangerous. Also, they have special utility in treating cancers that are inaccessible because of the amount of normal tissue that would be damaged in approaching them. Examples include treatment of pituitary cancers and/or destruction of pituitary as part of therapy, either for pituitary disease or for disease that pituitary excretions exacerbate.

Particles already in use either clinically or in clinical trials include negative pi-mesons, protons, and nuclei of atoms that have been stripped of their electrons. Clinical interest in negative pi-mesons is excited by their pattern of absorption in matter. The particle is intermediate in mass between electrons and protons and carries a unit negative charge. For much of its track it loses energy in a manner similar to electrons (low LET), but as it comes to rest, it may drop into the nucleus of one of the atoms of which living matter is composed. Added particles in atomic nuclei make them unstable, and they disintegrate. On so doing, massive, highly charged fragments fly apart from each other with considerable energy. In this manner the region occupied by a few cells is intensely irradiated with very high LET radiations. A great deal of destruction is to be anticipated as a result. In the initial portion of the pi-meson track there is no significant difference in the OER compared to photons (2.0 to 3.0), but in the final, very high LET component, RBE increases and the OER decreases. So far, there is little clinical experience with negative pi-mesons, and only a few facilities in the world produce them. It remains to be seen whether they will be superior to other modalities in controlling cancers, but for them to achieve widespread utility, they will have to be very superior indeed. They are expensive and complicated to produce and use and will need to be extremely good to justify their cost.

Proton beams are in use clinically. Their advantage over photons does not lie in reduction of the OER but in dose distribution. The beam can be restricted more accurately to well-circumscribed volumes than is possible using photons, thus enabling therapists to deliver higher doses to cancers while causing less damage to surrounding normal tissues. They will have their greatest utility, presumably, against cancers situated in or near radiosensitive normal tissue. These include various cancers of the abdomen and pelvis.

10.12 HYPOXIC CELL SENSITIZERS

Certain drugs have the property of imitating oxygen and, therefore, of sensitizing hypoxic cells. In practice these drugs are administered systemically and diffuse from capillaries into tissue. Unlike oxygen, they are not rapidly metabolized and can diffuse over considerable distances before being used up. Potentially, this gives them an advantage over hyperbaric oxygenation in sensitizing hypoxic cells. Added oxygen in capillaries diffuses from them into tissue but is very rapidly consumed. It is doubtful whether all hypoxic cells, particularly in cancers with many hypoxic cells, can reoxygenate using hyperbaric oxygen because of rapid oxygen consumption by cells nearest capillaries. Hypoxic cell sensitizers, on the other hand, may well sensitize all or nearly all hypoxic cells in cancers, because their slow rate of consumption prevents them from being entirely consumed by cells nearest capillaries.

Hypoxic cell sensitizers have an *affinity* for electrons, and this characteristic underlies their capacity for radiosensitization. Free radicals, as has already been stated, have an unpaired electron, and this electron is what makes them reactive and potentially damaging. The electron of free radicals and the electron affinity of hypoxic cell sensitizers make free radicals and these substances highly likely to react with each other. Such reactions may be represented as follows:

(1) $RH + \text{radiant energy} \longrightarrow R\cdot + H\cdot$
(2) $RH \longleftarrow R\cdot + H\cdot = \text{restoration}$
(3) $R\text{ (sensitizer)}\cdot + H\text{ (sensitizer)}\cdot$
   $= \text{no restoration}$

The organic molecule RH absorbs radiant energy and gives rise to the free radicals $R\cdot$ and $H\cdot$. In the absence of either oxygen or an electron-affinic agent, the free radicals $R\cdot$ and $H\cdot$ are drawn together and they react, restoring RH. However, in the presence of a sensitizer, R· and H· may react with the sensitizer. RH cannot be restored, and, to the extent that RH is important, damage is done. Clearly this mimics precisely the presumed action of oxygen (Section 10.7).

Medically useful electron affinic radiosensitizers must:

1. sensitize hypoxic cells but not be too toxic to normal tissue
2. be able to reach hypoxic cells (not be broken down by organs of the body or by well-aerated cancer cells)
3. be soluble in and able to diffuse through tissues (malignant or otherwise), because hypoxic cells in cancers are not near blood vessels
4. sensitize proliferating and/or quiescent cells since hypoxic cells are thought to be generally quiescent
5. be effective in the radiation dose ranges used in daily therapeutic fractions

Two compounds in the family nitroimidazole have stirred considerable interest. Metronidazole (commercial name, Flagyl) and misonidazole show promise in hypoxic cell sensitization both in experimental studies and in clinical trials. They demonstrate a good enhancement ratio (ER), a value used to quantify the action of radiosensitizing drugs.

ER is the ratio of *radiation doses* required to produce a given end point in given cell types in the presence and absence of radiosensitizers and is a measure of the sensitizing activity of the compound under consideration. Generally, various cell types at predetermined degrees of hypoxia are irradiated, and the results are compared to those obtained when the same cell types, fully aerated, are irradiated. This, of course, yields the *oxygen* enhancement ratio (OER). Then the same cell types at the same degree of hypoxia are given predetermined concentrations of electron-affinic agent and are irradiated. These results are compared to results of irradiation of the same cells when well aerated. This yields the ER and enables comparison to the OER.

Enhancement ratios have been worked out in vitro for metronidazole and misonidazole using cobalt 60 gamma rays and, depending on concentration of drug used, approximate 1.6 to 1.8. This is a considerable improvement over the use of cobalt 60 gamma rays in air, which yield an OER of about 3.0. Moreover, if these sensitizers are used with neutrons, there is an appreciable ER. Neutrons have, on their own, a lower OER than x or gamma rays—about 1.7 compared to 3.0 (Section 10.10)—and hypoxic cell sensitizers produce a smaller ER with them than they do with gamma rays. Nevertheless, there *is* an enhancement even with neutrons. Consequently, they may have potential for enhancing high-LET radiations to achieve perhaps even better therapeutic results.

The ER of metronidazole and misonidazole with gamma rays (about 1.6 to 1.8) is similar to the OER obtained with 14 MeV neutrons alone (about 1.7). In other words, based on in vitro experiments, the use of hypoxic cell sensitizers with low-LET radiations *could be* equivalent to using a higher LET radiation in radiation therapy.

The clinical performance of hypoxic cell sensitizers is still to be completely evaluated. Metronidazole has been tested in lower organisms (mice) and in human beings. In mice the dose of drug that could be given was small, but it did enhance the effect of low-LET radiations. In humans, results so far have also indicated that metronidazole produces some enhancement of the effects of low-LET radiations.

Before definitive evaluations are available of how useful these drugs will be clinically, they need to be tested against various cancers in various fraction patterns, in various doses of drug, and in conjunction with various radiations.

Misonidazole is a newer drug than metronidazole. As indicated, the amount of metronidazole tolerated by living organisms is small but does produce some enhancement. Misonidazole has a greater electron affinity than metronidazole, so that unit for unit of concentration, it would be expected to be more effective than metronidazole. That expectation has been met. Living test organisms tolerate only a small amount of it (about 1 μM), but that amount produces considerable sensitization. In the maximum achievable concentration, enhancement ratios approximating 1.8 are observed.

Effective use of hypoxic cell sensitizers leads to a potential problem. The drugs sensitize hypoxic cells in cancers and make it easier to kill them with radiation. However, with the dose levels used in daily fractions not *every* cell in treated cancers is killed. The effect of the sensitizer is to allow *given* radiation doses to kill *more* cells, but there will always be surviving cells. Sensitizers with radiations should shrink tumors more rapidly than radiations alone, and this probably permits more rapid and complete re-oxygenation of the *remaining* tumor mass. In turn, oxygenation permits previously hypoxic, quiescent cells to begin multiplying. When sensitizers are used, more cells should be killed by given radiation fractions than are killed without sensitizers, but more extensive *repopulation* of tumor may occur *between* fractions. It is *conceivable* that as much ground may be lost by repopulation between fractions as is gained by added cell kill. If this were to happen, spacing between fractions would be critical. Long periods between fractions permit re-oxygenation and repopulation, whereas short periods permit less re-oxygenation and repopulation. Presumably, an ideal spacing would exist, perhaps a different one for various tumor types and sizes, which would permit maximum re-oxygenation and maximum sensitization but would be too short for much proliferation and repopulation to occur. Possibly these ideal spacings could be determined reasonably precisely in animal models, but there is little likelihood that the kind of data needed to determine ideal fractionation

patterns can be gathered in humans. Experiments in lower animals can help demonstrate whether there are dramatic differences in tumor control using various fraction patterns. If there are, one might expect variable and perhaps unpredictable success clinically.

Experiments in lower animals designed to test this point suggest that spacing between radiation fractions *per se* is not without consequence—between fractions, regrowth may balance or nearly balance cell kill. When sensitizers are added to these experimental fraction schedules, it appears that, irrespective of the specific schedule, results are *always* improved. At least in these test systems, additional cell kill, owing to presence of sensitizers, does not lead to rampant repopulation. In turn, this suggests that the particular fraction pattern chosen for *clinical* therapy of humans may not critically affect the utility of hypoxic cell sensitizers. The data imply that any fraction pattern chosen to treat humans might benefit from using these sensitizers.

Apart from sensitizing hypoxic cells, electron-affinic sensitizers have a toxic effect on cells. For some reason they appear more toxic to hypoxic than to well-aerated cells. Because they kill hypoxic cells, they may help reduce the hypoxic cell component of tumors and may lead to more rapid shrinkage and re-oxygenation during radiotherapy.

It is clear from the foregoing that hypoxic cell sensitizers have considerable promise for use in clinical radiotherapy. Although much more work needs to be done to test them, it is already known that they cannot be used in unlimited quantities. They cause peripheral neuropathy; they can cause impairment of hearing; they may be carcinogenic; they certainly cause nausea, vomiting, and loss of appetite. Some of these dose-limiting effects can be mitigated by spacing of drug administration, but even so, there are side effects. Nevertheless, the study of such drugs is in a relatively early stage, and new electron-affinic agents may be produced which lack severe, dose-limiting side effects yet remain effective sensitizers.

10.13 RADIOPROTECTORS

It has been known since about 1950 that sulphydryl compounds (particularly the SH-groups) are capable of protection against tissue-damaging effects of irradiation. The mechanism of the effect is thought to be free-radical extinction, i.e., protection from free-radical damage. Sulphydryl compounds have been proposed as possible protectors of normal tissues during radiotherapy. This is based on the observation that these agents are capable of protecting normal tissues (in which they are widely and rapidly distributed) but not tumors in which, perhaps because of reduced blood supply, they distribute less well. Agents such as cysteine, cystamine, and glutathione have shown significant normal tissue protection but are rather toxic at the protective dose range. Newer agents such as ethylphosphorathionic acid (WR-2721)[16] and 2-mercaptopropionyl[17] have recently been proposed for this application because of their reduced toxicity at protective levels and their ability to increase radiation resistance by a factor of 2 to 3.

10.14 HYPERTHERMIA

In the last few years much interest and excitement have been elicited by encouraging radiotherapeutic results following irradiation of cancers raised to temperatures above body temperature (hyperthermia). Investigations into the efficacy of hyperthermia alone and of hyperthermia combined with low-LET radiations as potential cancer therapies have been carried out. Some of the biologic bases for the action of heat or heat combined with radiation have been established, and some of the parameters yet to be studied have been more clearly defined.

First, experiments carried out in tissue culture and in intact animals confirm that heating cells to temperatures not far above body temperature can kill, so heat by itself is, at least potentially, therapeutic. If cells are raised to temperatures somewhat above body temperature (usually given as 37.5° C) and maintained there, they begin to die. The longer they are kept warm, the greater the number of cells that are killed (Fig. 10.6).

Cells maintained at temperatures between 41° and 43° C exhibit a phenomenon called "thermotolerance." The *rate* of kill after a short time in this temperature range is *greater* than among those cells that survive long periods (see Fig. 10.6). It would seem that some cells are more sensitive to the same amount of heat than others and

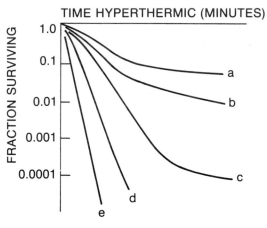

Fig. 10.6. A cluster of hypothetic curves describing cell survival as a function of time in hyperthermia. Progressing from curves a through e, each represents a higher temperature, but each differs from the one before or after it by only a small amount. For example, the type of relationship described by curve a is expected from mammalian cells kept at 41 to 42° C. Curves b, c, d, and e would be responses of the same cell types each maintained about 0.5° C higher than the one before it. Note that the response of cells kept at relatively low temperatures (curves a, b, and c) is biphasic—there is a change in slope. Evidently the *rate* of kill is less for cells kept long periods at temperatures *below* 43° C than for cells kept for shorter periods in the same temperature range. The phenomenon, called thermotolerance, is not exhibited at temperatures of 43° C and higher.

are killed early. The survivors tolerate that heat level.

Cells maintained at temperatures 43° C and above are killed at a more rapid rate than those maintained at lower temperatures, and they do not exhibit thermotolerance. The *mechanism* of cell killing by hyperthermia appears to be denaturation of cellular protein.

While hyperthermia alone can be used as cancer therapy, this is not based on preferential *cancer* cell killing. Cancer cells are not necessarily more heat sensitive than normal cells. It is the architecture of cancers which makes them especially vulnerable to the effects of heat. By the usual means of producing in vivo hyperthermia, both normal tissue and cancers are heated. However, the vasculature of normal tissue is well developed and efficient at carrying heat away. The vasculature of cancers is poorly developed and inadequate for the mass of cells it is required to service. This characteristic, besides causing hypoxia and nutrient deficiencies, also makes the vasculature inefficient at carrying off heat. Consequently, for the same amount of heat supplied over the same period of time, cancers become appreciably warmer than normal tissue. The differential can be quite important, since cell killing is more rapid and extensive at higher than at lower temperatures.

A most important factor is the fact that hypoxic cells are *not* resistant to heat. Heat sensitivity of hypoxic cells appears to be related to the fact that hypoxia results in low cellular pH, and cells at low pH are heat sensitive. Low pH may in turn result from buildup of lactic acid in hypoxic cancer cells which, in the absence of adequate oxygen, are compelled to be dependent on anaerobic metabolism. In any case, it is true that hypoxic cells are, at the least, not heat resistant and possibly even somewhat heat sensitive compared to normally aerated ones.

It is tempting to conclude on this basis alone that hyperthermia combined with low-LET radiations might be an effective means of cancer therapy. Radiations would be effective at killing well-aerated, proliferative cells, and hyperthermia at killing hypoxic, quiescent cells. Other considerations also point in the same direction. For example, of cells in proliferation, those in the latter portion of S are generally radioresistant compared to those in G_2, M, or even the early part of S. (See the next chapter for detailed discussion). Those same cells are sensitive to heat. Also, cells in G_1 may be radiation resistant, but they are heat sensitive.

The apparently complementary nature of hyperthermia and low-LET radiations strongly recommends the combination for clinical radiotherapy. Problems, however, remain to be solved before the method can achieve widespread use. One of these is the problem of local heating. In the usual course of events it is expected that the region of the body containing the cancer will be the only part heated. Only rarely will systemic hyperthermia be the goal. Immersion of body parts in warm water baths can be used, but it is effective principally for cancers in the extremities and for cancers at or near the body surface.

The bulk of cancers are deep-seated, and relatively few are in the extremities, so some kind of local, penetrating heat is desired. Shortwave diathermy, radiofrequency (RF) heat induction, microwaves, and ultrasound are all used. Unfortunately, all have shortcomings, making accurate, reproducible heat production to particular temperatures very difficult. Problems arise even in heat measurement. Thermometers of various kinds may warm up more than tissue, leading to inaccurate readings; they may disturb fields producing heat, resulting in uneven heating; or they can interfere with passage of energetic, heat-producing sound waves, producing some regions that are hot and some that are cooler. Some means of heat production (RF induction, for example) can lead to preferential heating of particular structures (e.g., bone).

Besides the problems mentioned above, the sequencing of heat and radiation in clinical settings needs to be worked out. In vivo animal studies show that the greatest cell kill for given radiation doses occurs when heat and radiation are given simultaneously. Clinically, that may not always be possible. In vivo animal studies further show that the closer together heat and radiation are given, the greater is the cell kill. It remains to be proved, however, that the same situation will obtain in humans and that a preferential effect will be exerted against cancers over normal tissue.

The problems remaining to be solved before hyperthermia can be used on a regular basis with low-LET radiations will not be solved quickly. There are various reasons. Producing accurate heat measurements is not a simple task. New heat measuring materials may have to be invented before this can be done, and that will take time. Exploration of such questions as what kind of heating is best in given circumstances, what kind of complications and rates of complications will occur, how the combined modalities will affect re-oxygenation and repopulation, and whether or not heat can be used with combined radiation-chemotherapy regimens will be very time-consuming. Still, the experimental data are encouraging and lead to the hope that hyperthermia will be an important tool in the hands of radiotherapists.

10.15 DOSE RATE

The biologic effects or degree of effectiveness of given doses of many radiations is dependent on the *rate* at which the radiation is given. The influence of dose rate is very significant for low-LET radiations but is not observed with very high LET radiations. Dose-rate influence really is a reflection of repair of sublethal cellular damage. Since very high LET radiations produce practically only lethal cellular damage, there is no influence of dose rate on their effects. Whenever a very high LET radiation passes through a cellular nucleus, the chances are nearly 100% that it will kill the cell. Consequently, the number of cells killed by very high LET radiations depends only on the number of such radiations that pass through nuclei of cells in an irradiated population. Whether these radiations are delivered rapidly or slowly is irrelevant.

Very low LET radiations passing through cellular nuclei are not invariably lethal. In fact, when they cause injury, the damage they do is most likely to be *sub*lethal. Sublethal damage apparently is reparable, because sublethally injured cells recover. The following chapter deals with this subject in detail. The repair of sublethal radiation damage takes time, but given enough time (and other conditions permitting), full repair occurs. Whether low-LET radiations are delivered slowly or rapidly, the same number of sublethal injuries doubtless occur. However, when radiations are delivered very slowly, there is time for each sublethal injury to be repaired before the next one occurs. If the *rate* of radiation delivery is increased, the amount of time between each sublethal injury decreases, and the probability of repair grows less. At rapid rates of radiation delivery, sublethal injuries are likely to accumulate unrepaired and to reach lethal levels. At very low rates of radiation delivery they do not. Particular radiation doses will, therefore, produce more cell killing when radiation is delivered rapidly than when it is delivered slowly.

Survival curves obtained following cellular irradiation with very low LET radiations at different dose rates reflect this phenomenon (Fig. 10.7). Survival curves using low-LET radiations at very low rates are exponential, indicating that no sublethal damage is registered. Sublethal damage may have been done, but all of it is repaired; none can accumulate and none is

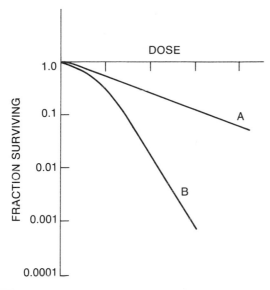

Fig. 10.7. Hypothetic mammalian cell-survival curves illustrating the influence of dose rate on cell survival using low-LET radiations (x or gamma rays). Curve A is the response expected when the rate of radiation delivery is slow. The slope of the curve is quite shallow, and the curve itself is fully exponential (n = 1). Curve B typifies response when dose rate is rapid. Its slope is steeper and there is a shoulder (n > 1).

detected. Cell number, however, *is* reduced; cells are killed. These cells are, in all probability, those few which were near or at the ends of the tracks of the low-LET radiations. Track tails are short and comprise a small fraction of any track, and this probably accounts for the observation that few cells are killed by any unit of radiation dose (the curve's slope is shallow; D_o is very large). At extremely low dose rates, the number of cells in an irradiated population may not decline at all or may even increase. This happens when the rate of cell proliferation in the irradiated population is as great or greater than the rate of killing of cells in the tails of low-LET tracks. Something like this must happen to cell populations irradiated by the background. Background radiation must occasionally kill cells. Yet, cell numbers may increase (embryos go from one cell to enormous numbers, and children grow to adults), or the numbers may remain constant (cell numbers in adults remain more or less constant).

At higher dose rates, survival curves have an appreciable shoulder, and the slope is steeper (see Fig. 10.7, curve B). The existence of a shoulder means that n is greater than 1.0, an indication that lethal levels of injury are being achieved by the accumulation of several sublethal events. The steeper slope results from the fact that many more cells are now vulnerable to killing by given quantities of radiation dose. Not only will cells at the tails of tracks be killed, but others, not at the tails, that receive sublethal injuries too close together in time to be repaired, may also be killed.

10.16   CLINICAL IMPLICATIONS

Dose-rate effects probably play no important role in diagnostic radiology using x rays. Generally, exposures are made in fractions of seconds, dose rates that, by most standards, would have to be regarded as rapid. As stated many times, at the *dose* levels used in diagnostic radiology, lethal cell damage probably does not occur in a significant number of cells, and the production of the sublethal cellular effects, carcinogenesis and mutation—at diagnostic dose levels—is unlikely to be dose-rate dependent.

The dose-rate effect is potentially exploitable in radiation therapy in an interesting way. If the potential for repair of or recovery from sublethal damage for certain cancers and for normal tissues is different, and if the difference is in favor of normal tissues, dose rate may be used to possible advantage. For instance, cells of gastrointestinal tract are relatively radiosensitive; values of D_o are relatively low. Yet, these same cells have good potential for repairing sublethal injuries; curve shoulders are substantial. Generally, cancers that have spread to and located themselves near or on gastrointestinal tissue present difficulties in treatment because of the tissue's dose limitations. However, if a slow dose rate could be used in treatment, gastrointestinal tissue might be able to repair much of the damage sustained during treatment exposure, but the cancer, if its repair potential were less, would be able to repair less, and it would accumulate lethal damage levels. "Sparing" of normal tissue, as well as better tumor control, might result. Work in animal models does suggest that benefits can result from manipulation of dose rate, but there are too few clinical data to establish the point in humans.

SUMMARY

1. The probability that cells will survive any given radiation dose depends on the LET of radiations used. Low-LET radiations are relatively less effective than high-LET radiations.
2. Oxygen is a radiosensitizer. Under conditions of moderate or short-term hypoxia, D_o is increased but n is unchanged. The effect is most pronounced with low-LET radiations and tends to diminish with high-LET radiations. Under severe or long-term hypoxia, D_o is increased, but n also changes, approaching 1.0.
3. Altering dose rate may markedly change cellular response to radiations. Sublethal damage may be repaired during irradiation at low dose rates, but it accumulates during irradiation at high dose rates.

REFERENCES

1. Humphrey, R.M., Dewey, W.C., and Cork, A.: Effect of oxygen in mammalian cells sensitized to radiation by incorporation of 5-bromodeoxyuridine into the DNA. Nature, *198*:268–269, 1963.
2. Revesz, L., and Littbrand, B.: Variation of the relative sensitivity of closely related neoplastic cell lines irradiated in culture in the presence or absence of oxygen. Nature, *203*:742–744, 1964.
3. Belli, J.A., Dicus, G.J., and Bonte, F.J.: Radiation response of mammalian tumor cells: I. Repair of sublethal damage in vivo. J. Natl. Cancer Inst., *38*:673–682, 1967.
4. Berry, R.J.: Some observations on the combined effects of x-rays and methotrexate on human tumor cells in vitro with possible relevance to their most useful combination in radiotherapy. Am. J. Roent., *102*:509–518, 1968.
5. Elkind, M.M., Withers, H.R., and Belli, J.A.: *In* Frontiers of Radiation Therapy, 3rd Ed. Edited by J.M. Vaeth. Basel, S. Karger, 1968.
6. Phillips, T.L.: Qualitative alteration in radiation injury under hypoxic conditions. Radiology, *91*:529–536, 1968.
7. Van Putten, L.M., and Kallman, R.F.: Oxygenation status of a transplantable tumor during fractionated radiation therapy. J. Natl. Cancer Inst., *40*:441–451, 1968.
8. Elkind, M.M., Withers, H.R., and Belli, J.A.: Intracellular repair and the oxygen effect in radiobiology and radiotherapy. Front. Radiat. Ther. Oncol., *3*:55–87, 1967.
9. Nias, A.H.W., Swallow, A.J., Keene, J.P., and Hodgson, B.W.: Absence of a fractionation effect in irradiated Hela cells. Int. J. Radiat. Biol., *23*:559–569, 1973.
10. Thomlinson, R.H., and Gray, L.H.: The histological structure of some human lung cancers and the possible implications for radiotherapy Br. J. Cancer, *9*:539–549, 1955.
11. Warburg, O.: The Metabolism of Tumors. London, Constable, 1930, p. 6.
12. Uritasun, R.C., et al.: Radiation and high dose metronidazole (Flagyl) in supratentorial glioblastomas. N. Engl. J. Med., *294*:1364–1367, 1976.
13. Kallman, R.F., Jardine, L.J., and Johnson, C.W.: Effects of different schedules of dose fractionation on the oxygenation status of a transplantable mouse sarcoma. J. Natl. Cancer Inst., *44*:369–377, 1970.
14. Churchill-Davidson, I., Sanger, C., and Thomlinson, R.H.: Oxygenation in radiotherapy. II. Clinical application. Br. J. Radiol., *30*:406–442, 1957.
15. Churchill-Davidson, I.: Hyperbaric oxygenation. Ann. NY Acad. Sci., *117*:647–890, 1965.
16. Yuhas, J.M.: A more general role for WR-2721 in cancer therapy. Br. J. Cancer, *41*:832–834, 1980.
17. Nagata, H.: Studies on sulfhydryl radioprotectors with low toxicities. Tokushima J. Exp. Med., *27*:15–21, 1980.

Chapter 11

Repair and Recovery from Radiation Damage

11.1 INTRODUCTION

The sparing effect of low dose rates of low-LET radiations suggests recovery from sublethal cellular injury is possible. A great deal of experimental work, using techniques other than variable dose rates, also demonstrates such recovery.

11.2 POTENTIALLY LETHAL AND SUBLETHAL DAMAGE

Two types of nonlethal cellular damage apparently occur, called *potentially lethal* and *sublethal,* respectively. At the moment, it is not certain whether these are distinct or only apparently distinct injuries, but considerable evidence indicates that they are different.

Sublethal cellular damage has been described in some detail in previous chapters; it is the form of damage thought to be responsible for shoulders on survival curves. The existence of this form of damage is *inferred* from the shape of survival curves, but so far it has not been directly demonstrable. Survival curves are a graphic representation of the relationship between the fraction of cells surviving and being killed by given radiation doses. Survival curves,

therefore, are a record of the production of *lethal* damage. Sublethal damage is presumed to occur because of the steepened slope in the dose range beyond the shoulder. The change in slope shows increased vulnerability to radiation at higher doses, suggesting an accumulation of sublethal injuries has occurred in the low-dose range and is reaching lethal levels in the high-dose range. The sublethal injuries themselves are not detected; the only time anything is noted is when a cell is lethally injured and dies.

Potentially lethal damage (PLD) is damage which, under some circumstances, is lethal but under others, is not. It appears to be an injury which may, under the right conditions, be expressed and detected. Individual sublethal injuries are never lethal and apparently have no potential for being lethal. Potentially lethal damage, under the right conditions, can be lethal. Other evidence also points to differences between sublethal and potentially lethal lesions. For example, repair of potentially lethal damage is more rapid than repair of sublethal damage;[1] conditions that produce reduced capacity for repair of potentially lethal damage do not necessarily result in loss of capacity to repair sublethal lesions.[2] Nevertheless, the nature of these forms of damage is still incompletely understood and there remains the possibility that they are not inherently different, but only different manifestations of the same phenomenon.

11.3 POTENTIALLY LETHAL DAMAGE (PLD)

If single doses of low-LET radiations are given to two identical cell populations under identical conditions, the probability of survival will differ according to *post*-irradiation conditions. If one of the populations is actively proliferating and the other is quiescent, the fraction of cells that survives will be appreciably smaller in the proliferating population than in the quiescent one. Such observations have led to the supposition that a form of injury or injuries can occur which causes death in cells that are actively replicating but which is not lethal in cells that are quiescent. Presumably the potentially lethal injury is produced both in proliferating and quiescent cells, but quiescence is a condition which favors *repair* of this injury.[3-6]

11.4 SUBLETHAL DAMAGE

Sublethal damage is not expressed except when enough of it has been accumulated to cause cell death. Recovery is demonstrated by split-dose experiments. Briefly, a cell population is exposed to sufficient radiation so that any survivors will have accumulated the maximum amount of sublethal damage. A time interval is permitted to elapse. The survivors are irradiated again. In the absence of any recovery, the proportion of irradiated cells expected to survive the two radiation exposures can be easily predicted; it will be the same as would survive a *single* radiation exposure equal to the size of the two exposures combined. When experiments are carried out, the *observed* proportion of survivors of the two exposures can be greater than this prediction. The explanation usually given is

that, in the time lapse between exposures one and two, cells that survive exposure one recover from damage done during that exposure. Radiation given in a single exposure includes no time lapse and no time for recovery, so there will be fewer survivors (Fig. 11.1). During recovery, the shoulder of the survival curve reappears, whereas when the first exposure is concluded, maximum sublethal damage has been done (n = 1). The size of the fraction surviving the second exposure indicates that, during the time lapse between exposures, cells acquire again the capacity to absorb sublethal damage (n becomes greater than 1). Moreover, the average number of hits (n) in unirradiated popu-

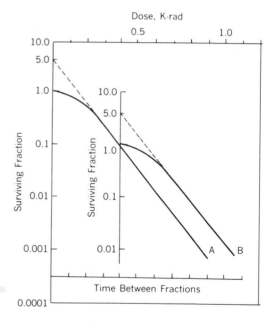

Fig. 11.1. Results of a hypothetic split-dose experiment involving irradiation of mammalian cells where enough time for full recovery is permitted between fractions. The first fraction (called the *conditioning dose*) in this example is about 400 rads, but the second varies in magnitude.

Curve A shows the response when given radiation doses are delivered in single exposures. Curve B shows the results when the same doses are split into two fractions. A dose, for example, of 800 rads given at once produces a survival level of about 0.001 (Curve A), but when split into two fractions, a survival level nearer to 0.01 is observed (Curve B). Extrapolation of both Curves A and B gives the same extrapolation number.

lations and in recovered populations is the same. This means that recovery is complete. No residual, unrecovered or unhealed damage is detected, at least none reflecting itself in a change of extrapolation number.

Recovery from sublethal damage eventually becomes complete, but it takes time. When mammalian cells are subjected to two radiation exposures separated by *variable* time intervals, the surviving fraction increases, reaches maximum, declines and then increases again (Fig. 11.2).

The *initial* rise represents recovery from sublethal damage. Presumably, sublethal damage caused by the first exposure (called the conditioning dose) is repaired. The more time that passes before a second dose is administered, the more damage is repaired and the more complete the recovery. The amount of time required for *full* recovery from maximum degrees of sublethal damage varies, but may be as much as 4 to 5 hours,[7] while repair of potentially lethal damage takes only about 1 hour.

The drop in surviving fraction which follows the initial rise is not related to recovery or repair. That is due to synchronization of cells in life cycle by the conditioning dose and irradiation of synchronously dividing cells during sensitive life-cycle phases.

11.5 RADIOSENSITIVITY AND CELL LIFE CYCLE

Radiosensitivity has been defined as dose of radiations required to produce hits or inactivate targets. The property, radiosensitivity, is not the same in all phases of a cell's life; hit production apparently requires a greater or lesser radiation dose, depending on when in the cycle radiation is given.

Life cycles of cells are divisible into four recognizable stages (Fig. 11.3). The first of these is G_1, *interphase*, the stage between reproductive episodes. While not reproductive *per se*, activities occur in G_1 which

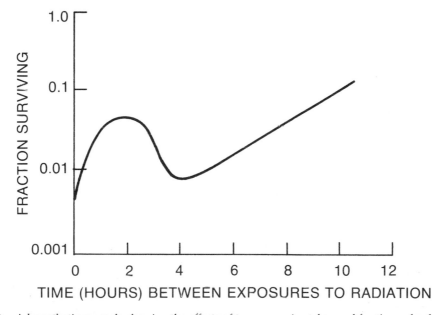

Fig. 11.2.   A hypothetic example showing the effects of two approximately equal fractions of radiation dose, separated by various time intervals, on mammalian cell survival. When both dose fractions are given together (time zero), the surviving fraction of cells is small. As time between dose fractions increases, surviving fractions of cells also increase. A peak in surviving fraction is followed by a reduction in this end point which, in turn, is followed by another rise.

are necessary prerequisites for cells' entry into the next phase, S. Evidently a form of RNA is synthesized during G_1 in cells which are to reproduce, and this RNA is necessary before S can successfully commence.[8,9]

Cell reproduction or replication begins with the phase called S. During S, new DNA is synthesized, and when S concludes, a short phase, called G_2, begins. In G_2 cells synthesize certain protein and RNA molecules necessary for entry into and successful completion of the next and last phase, mitosis (M).[10] In mitosis, cells, having replicated DNA and chromosomes, divide to produce two cells from one.

The length of time required for the reproductive phases S, G_2 and M does not vary much among mammalian cells. Once mammalian cells begin reproduction they pass through these phases in roughly the same time. It is the time between reproductive episodes (G_1) that varies.

Variations in radiosensitivity as a function of position of cells in life cycle differ according to the length of G_1. In cells with short G_1 periods, mitosis (M) is most sensitive. Progressing from M, G_1 is more resistant, and resistance steadily increases during this phase, continues to increase in S, and reaches a peak in late S. There is a steep decline during G_2, reaching a low again in M.[11] In cell types in which G_1 is long, the pattern is similar from the *end* of G_1 onward toward mitosis; however, there is, in addition to the peak of resistance in S, a resistant phase in early G_1 (Fig. 11.4).

In tissue cultures or in proliferating compartments of tissues and organs, there generally are cells in all phases of life cycle.

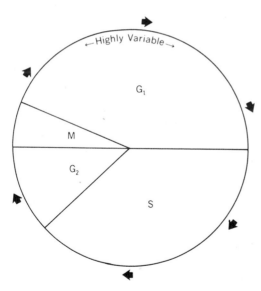

Fig. 11.3. A diagrammatic representation of cellular life cycle. It begins after M, mitosis, so the earliest stage is G_1. The time in G_1 is highly variable, probably ranging from a few minutes to many months depending on cell type. Cells are thought of as "aging" as they progress through the cycle. S, the stage in which DNA is synthesized, is usually long compared to either G_1 or G_2, and M is usually very brief. These approximate proportions are typical of many mammalian cells, but individual types may deviate. The symbol G stands for the word *gap*, because the processes going on in G_1 and G_2 are not completely known and are gaps in knowledge.

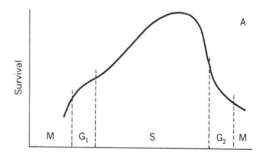

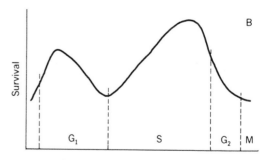

Fig. 11.4. Illustrations depicting typical responses of proliferative capacity of mammalian cells following given doses of radiation. *A* shows the response of cells having short G_1 periods and *B* those having long G_1 periods. With some exceptions, these appear typical of most mammalian cell lines tested. For example, in *A*, cells irradiated in S are most likely to survive, in M, least likely. Note that differences in survival levels between high and low points is a factor of nearly three, a common observation in experimental systems.

Because of this, given doses of radiation will affect cells in the various phases to different degrees. Those in sensitive phases will be more severely damaged than those in more resistant phases. In any phase, surviving cells with sublethal radiation damage may repair their damage, but those with the greatest degree of damage will take the most time to recover from it. Consequently, cells in S will incur the least damage from any given dose, because they are least sensitive and can repair or recover from damage most quickly. Whereas proliferating cells in tissues and tissue cultures usually are in all stages of life cycle, the effect of an exposure to radiation is to synchronize them in a single phase. Cells in resistant phases, least damaged and most rapidly recovered, are first to resume normal activity (viz, proliferation) after irradiation. Those in sensitive phases are last to resume proliferation. They are most damaged and need the most time to recover. This phenomenon explains the shape of the curve in Fig. 11.2. The initial rise represents recovery of cells that repair sublethal damage first, that is, the most resistant and least damaged, those in late S. These cells can and do resume progression through proliferative stages, because they have recovered. They are referred to as a *cohort*, cells, all in the same phase when irradiated (S), progressing together after recovery through the same life-cycle phases, G_2 and M. G_2 and M are, compared to S, radiosensitive. Thus, a second radiation exposure after the *synchronizing*, conditioning dose, can greatly reduce the surviving fraction, since the entire proliferating cohort may be in the same radiosensitive phases, G_2 and M. The *continued* rise in surviving fraction, observed when the second exposure is given many hours after the conditioning dose (see Fig. 11.2), occurs because by then, most cells in the cohort have passed through M, and have divided into two, multiplying the population being exposed.

Variations in radiosensitivity in different phases of life cycle seemingly are due in part to DNA synthesis itself. Survival after irradiation with low-LET radiations depends *intimately* on the onset of DNA synthesis or some process concomitant with it (see Fig. 11.4). Survival rises as cells leave G_1 and begin synthesizing DNA.[12,13] However, at least one other factor appears to exert a controlling influence on the cyclic dependence of lethal radiation damage in mammalian cells. This putative factor has been called "Q factor," and it is believed to either protect cells from radiation damage or to repair radiation damage before it can be expressed. As has been mentioned in previous chapters, the nature of Q is unknown, but studies suggest it may be identified with a portion of the intracellular sulfhydryl.

Since proliferating cells need time to repair radiation damage, irradiation causes a delay in progression through life cycle. During the delay, damage is repaired, and when cells have fully recovered, they resume proliferation.

Variations in sensitivity related to position in life cycle occur mainly when cells are irradiated with low-LET radiations; the nature of damage done by high-LET radiations is such that position in life cycle is a lesser factor in survival and there are much smaller variations in sensitivity to these radiations.

Finally, variations in sensitivity to radiation during life cycle are not related to possible differences in oxygen concentration. Oxygen enhancement is constant for all phases of life cycle.

11.6 REPAIR OF SUBLETHAL DAMAGE

Repair of sublethal damage, however it may occur, evidently is independent of post-irradiation conditions but is dependent on oxygen tension. In conditions of hypoxia, this form of damage is either irreparable or repaired to a very minor degree. Hypoxia thus presents a peculiar but important paradox. Hypoxic cells are radio-resistant; D_o in them is higher than in normally aerated cells of the same kind. More radiation energy must—on the average—be expended to produce hits in hypoxic cells than in normally oxygenated ones. However, *because* hypoxia prevents repair of sublethal damage, whatever sublethal injuries are produced remain unrepaired for as long as cells remain hypoxic. If well-aerated and hypoxic cells of the same kind are given the same dose of low-LET radiations, well-aerated cells will be more severely damaged, but survivors will repair and recover from that damage. Hypoxic cells will be less severely damaged, but survivors will not repair and recover. In radiotherapy regimens consisting of a number of fractions, normally aerated cells that survive each fraction repair and recover from injuries produced in each fraction. Injuries do not accumulate from fraction to fraction. Hypoxic cells that do not re-oxygenate do not repair and recover from injuries produced in each fraction. These add to each other, eventually perhaps, to reach lethal levels.

Hypoxia prevents repair of sublethal injuries, but in a sense it favors repair of potentially lethal injuries. Hypoxic cells cannot proliferate, and quiescence is a condition that favors repair of potentially lethal lesions. This and the fact that most hypoxic cancer cells probably re-oxygenate at some point during fractioned radiotherapy regimens, add to the hazard presented to cancer patients by hypoxic cancer cells. Potentially lethal damage produced by given radiation doses is expressed more often in normally aerated cells than in hypoxic ones. Upon re-oxygenation and re-population, hypoxic cells will have repaired potentially lethal damage and may then multiply to form tumor recurrences.

11.7   THE QUASI-THRESHOLD (D_q)

The shoulder of survival curves can be regarded as a kind of threshold. Radiation doses less than those that bring cell populations to the shoulder probably produce principally sublethal cellular damage, but doses greater than those that bring cell populations to the shoulder produce enough damage to be lethal. The shoulder is a quasi-threshold; from doses below it mainly sublethal damage occurs, above it mainly lethal damage occurs. The exact location of the shoulders on survival curves is not always accurately known but the quasi-threshold *concept* is so useful that its location on survival curves has been defined. The quasi-threshold (D_q) is the intercept of an extrapolation, beyond the shoulder, of the exponential portion of survival curves with 100% survival. It is the dose of radiation beneath which mainly sublethal, reparable damage is caused and above which lethal, irreparable damage is produced.

The utility of D_q is a measure of repair of sublethal damage, and is especially useful in biologic systems where the shoulder of the curve is not directly measurable. In tissue cultures, surviving fractions of cells are known because, in any experiment, the number of cells irradiated is easily determined. However, the same cannot be done when trying to construct survival curves in vivo. No one knows how many cells there are in any given portion of any given tissue and/or how many are proliferative. Because the starting number of cells is unknown, the surviving fraction of any radiation dose cannot be known either. Since this is so, the position of the shoulder cannot be determined, but production and repair of sublethal damage in vivo can be studied using D_q.

A commonly used technique is as follows. A section of tissue to be evaluated in vivo is marked off. For example, a section of skin of certain dimensions on the backs of mice may be used, or predetermined lengths of intestine in test organisms may be selected. Within the section of skin or length of gut, an area is set aside to determine the effects of *test* doses of radiation, and the region *surrounding* this test area is given a large radiation dose, sufficient to sterilize all cells in it. The test area is then given a predetermined dose of radiation, one that will be lethal to many but not all cells in the zone. Eventually, within the test area, colonies of cells will begin to appear. These are descendants of cells in the zone that survived the test radiation dose with their ability to reproduce intact. The purpose of the sterile zone is to assure that any colonies observed *in* the test zone originate from cells there and are not the result of migration of unirradiated cells from adjacent tissue.

Using a number of test zones and test radiation doses, and determining the number of colonies arising after each test dose, a curve can be plotted. What is plotted is *not* surviving fraction since this is unknown, but *number* of *colonies* per unit of tissue tested (e.g., square centimeter of skin) as a function of radiation dose. To be reasonably certain that each observed colony is derived from one surviving cell, the test radiation doses must be high, high enough so that there are a few survivors and that these survivors are far between. The result is that most, if not all, surviving cells have absorbed maximal degrees of sublethal damage, and only the final exponential portion of the survival curve is obtained. The shape of the curve before this exponential is not known. However, it is possible to determine what radiation dose produces maximal reparable damage; that is, it is possible to determine D_q which is, after all, a measure of the *width* of the shoulder.

A radiation dose (Fig. 11.5) is given *in a single exposure* that yields a certain number of colonies; for example, a dose might be given that yields five colonies per square centimeter of skin. Then, the dose given *in two exposures*, separated by a time lapse, needed to produce the *same* number of colonies in the same unit of tissue is determined. Because of the repair of sublethal damage, the latter will be greater than the former. The *difference* between the single dose and the total of the split doses required to yield the same number of colonies will be D_q, the dose that produced sublethal radiation damage which was repaired in the time lapse.

The use of this method permits survival curves to be constructed from data collected in vivo and comparisons to be made with curves obtained from the same kinds of cells irradiated in vitro. Results show that the *shapes* of both curves are similar and indicate that in vitro data are a valid model for studying radiation effects on cells in tissues.

11.8   TARGET OF SUBLETHAL AND POTENTIALLY LETHAL DAMAGE

Sublethal and potentially lethal damage appear to be different things. It is not clear, however, whether they represent radiation injury to different cellular structures or different *kinds* of injury to the same cellular structure.

Much evidence supports the contention that DNA is the principal target for killing of mammalian cells by ionizing radiations.[14][16] What is unclear is what the principal lesion in DNA responsible for this effect might be and whether sublethal and potentially lethal damage are different manifestations of this lesion, different lesions in DNA molecules, or lesions in different molecules altogether. Evidence points to the possibility that potentially lethal damage may consist of single-strand DNA breaks,[17,18] but sublethal damage may consist of damage at a more complex level, perhaps at the level of both DNA strands, which, when unrepaired, causes enough distortion of the DNA-protein complex to lead to chromosome damage and cell death.[19–22]

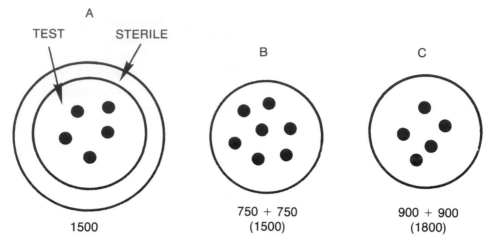

Fig. 11.5. The technique for in vivo measurement of D_q. In A, a single dose of 1500 rads yields 5 colonies in a test zone of specified dimensions. When this dose is divided into two equal fractions (B) separated by time, more (7) colonies appear in a zone of the same dimensions. To achieve a 5-colony survival with 2 doses (C) requires 1800 rads. The difference in dose between A and C, 300 rads, owes to repair of sublethal damage and is D_q.

11.9　POTENTIALLY LETHAL DAMAGE AND MALIGNANT TRANSFORMATION

Certainly there should be no need to repeat here that exposure to ionizing radiation in vivo is carcinogenic. It may not, however, be as well known that nonmalignant cells in tissue culture may, after exposure to carcinogens, including ionizing radiations, display properties associated with cultured malignant cells. Moreover, some of these cells, when implanted into animals, give rise to tumors. Carcinogens, therefore, seem to be able to produce malignant transformations in vitro, and this is a good system in which to study various aspects of carcinogenicity.

There are reports linking repair of potentially lethal radiation damage (PLD) and enhancement of radiation-produced malignant transformation.[23] A condition favoring repair of potentially lethal damage in tissue culture is "density inhibition." Commonly, if a tissue culture vessel has so many cells in it that they touch, proliferation ceases. The ability to proliferate is not lost or destroyed, because on removing some

of these quiescent cells to an empty culture vessel, proliferation resumes. Provided there is sufficient nutrient, proliferation will continue until this vessel also is densely populated; then proliferation will cease again.

Survival from given doses of radiation is increased if mitosis is prevented by density inhibition for two to four hours after irradiation. This is presumed to result from repair of potentially lethal damage.[5,24] Concomitant with increased survival, the frequency of malignant transformation is enhanced;[23] the *more* survival is increased by keeping culture density inhibited for time periods up to four hours, the higher is the frequency of malignant transformation. However, if mitosis is prevented for *longer* than two to four hours, the frequency of malignant transformation begins to decrease, declining until, after 24 to 48 hours of density inhibition, the frequency is less than that of proliferating cells, that is, those that do not get the chance to repair potentially lethal damage at all.

A hypothesis has been advanced to explain these data. Potentially lethal damage is believed to be injury to DNA. (This is of course in keeping with the hypothesis, presented in the previous section, that potentially lethal damage may be single-strand DNA breaks.) During repair of radiation-induced DNA lesions, errors are inserted in the DNA, and these lead to malignant transformation. This accounts for increase in frequency of malignant transformation as potentially lethal damage is repaired. However, if, when potentially lethal damage *is* repaired, cells continue to be held in mitotic inhibition, they may then go back over their DNA and delete and correct the errors made during repair of potentially lethal injuries. The degree to which erroneous repair of DNA can be corrected depends on when cell division occurs; the longer the density inhibition, the more time for error correction and the lesser the transformation frequency.

The phenomenon of enhanced transformation with repair of potentially lethal damage is established in vitro, and if it applies in vivo, there are the following implications. Should exposure to carcinogens in vivo be associated with an appropriate proliferative stimulus, slowly proliferating or nonactively cycling cell populations, which are frequent targets of environmental carcinogens, may be particularly vulnerable to malignant transformation. An example is in order. Suppose that a tissue consisting of both proliferating and nonproliferating cells is exposed to the carcinogen, ionizing radiation. Potentially lethal damage will be produced in its cells. It will not be repaired in many injured proliferative cells and these will die. It may be repaired in quiescent cells. These will live, but for a time there will be a high incidence of malignant transformation. Death of the proliferative cells causes a reduction in cell number in the tissue and a proliferative command is received to repopulate the tissue. Quiescent cells begin proliferating in response, and among them will be the *relatively* large population of transformed cells.

SUMMARY

1. Two kinds of nonlethal cellular lesions are produced by radiation, potentially lethal and sublethal.
2. Potentially lethal lesions are repaired in nonproliferating cells but not in replicating ones. Repair is rapid, occurring in about one hour, and the lesion is likely to be single-strand DNA breaks.
3. Repair of potentially lethal damage is associated with enhancement of malignant transformation in mammalian cells.
4. Repair of sublethal damage is slower than that of potentially lethal lesions and is dependent on sufficient oxygenation. Sublethal damage is believed to result from lesions in DNA of a more complex order than single-strand breaks.
5. Radiation sensitivity, particularly to low-LET radiations, depends on stage in life cycle. Differences between the most sensitive (M) and least sensitive (late S) stages may be a factor of three. Changes in sensitivity are intimately associated with DNA synthesis or a process concomitant with it, but are also controlled by an as yet unidentified substance, Q, which may be associated with cellular sulfhydryl.
6. Cell-cycle variations in radiosensitivity are not associated with oxygen concentrations; oxygen enhancement is uniform throughout the cycle.
7. D_q denotes the quasi-threshold radiation dose. Beneath this dose, principally sublethal damage is done. Above it, lethal damage occurs. It is the dose causing maximal sublethal, reparable damage.

REFERENCES

1. Elkind, M.M., and Redpath, J.L.: Molecular and cellular biology of radiation lethality. *In* Cancer:

A Comprehensive Treatise. Vol. 6. Edited by F.F. Becker. New York, Plenum Publ. Corp., 1977, pp. 51–59.

2. Utsumi, M., and Elkind, M.M.: Potentially lethal damage versus sublethal damage: Independent repair processes in actively growing Chinese hamster cells. Radiat. Res., 77:346–360, 1979.

3. Phillips, R.A., and Tolmach, L.J.: Repair of potentially lethal damage in x-irradiated Hela cells. Radiat. Res., 29:413–432, 1966.

4. Belli, J.A., and Shelton, M.: Repair by mammalian cells in culture. Science, 165:490–492, 1969.

5. Little, J.B.: Repair of sublethal and potentially lethal radiation damage in plateau phase cultures of human cells. Nature, 224:804–806, 1969.

6. Hahn, G.M., and Little, J.B.: Plateau-phase cultures of mammalian cells: An in vitro model for human cancer. Curr. Top. Radiat. Res., 8:39–83, 1972.

7. Elkind, M.M.: Cells, targets, and molecules in radiation biology. In Radiation Biology in Cancer Research. Edited by R.E. Meyer and H.R. Withers. New York, Raven Press, 1980.

8. Baserga, R., Estensen, R.D., and Peterson, R.O.: Delayed inhibition of DNA synthesis in mouse jejunum by low doses of actinomycin-D. J. Cell Physiol., 68:177–184, 1966.

9. Baserga, R.: Multiplication and Division in Mammalian Cells. New York, Marcel Dekker, Inc., 1976, pp. 32–34.

10. Sinclair, W.K.: Cell cycle dependence of the lethal radiation response in mammalian cells. Curr. Top. Radiat. Res., Quarterly I: 264–285, 1972.

11. Teresima, T., and Tolmach, L.J.: X-ray sensitivity and DNA synthesis in synchronous populations of Hela cells. Science, 140:490–492, 1963.

12. Sinclair, W.K.: N-ethylmaleimide and the cyclic response to x-rays of synchronous Chinese hamster cells. Radiat. Res., 55:41–57, 1973.

13. Sinclair, W.K.: Dependence of radiosensitivity upon cell age. In Time and Dose Relationships as Applied to Radiotherapy. Brookhaven National Laboratory Report BNL-50203 (C-57), NCI-AEC Conference, Carmel, California, 1970, pp. 97–116.

14. Cole, A., et al.: Mechanisms of cell injury. In Radiation Biology in Cancer Research. Edited by R.E. Meyer and H.R. Withers. New York, Raven Press, 1980, pp. 33–58.

15. Painter, R.B.: The role of DNA damage and repair in cell killing by ionizing radiation. In Radiation Biology in Cancer Research. Edited by R.E. Meyer and H.R. Withers. New York, Raven Press, 1980, pp. 59–68.

16. Ben-Hur, E., and Elkind, M.M.: Damage and repair of DNA in 5-bromodeoxyuridine labelled Chinese hamster cells exposed to fluorescent light. Biophys. J., 23:336–347, 1972.

17. Man, A., and Elkind, M.M.: Additive action of ionizing and non-ionizing radiations through the Chinese hamster cell cycle. Int. J. Radiat. Biol., 31:275–282, 1977.

18. Sasaki, M.S., and Norman, M.: Selection against chromosome aberrations in human lymphocytes. Nature, 214:502–503, 1967.

19. Carrarro, A.V.: Chromosome aberrations and radiation induced cell death. I. Transmission and survival parameters of aberrations. Mutat. Res., 17:341–353, 1973.

20. Carrarro, A.V.: Chromosome aberrations and radiation induced cell death. II. Predicted and observed cell survival. Mutat. Res., 17:355–366, 1973.

21. Bedford, J.S., Mitchell, J.B., Griggs, H.G., and Bender, M.A.: Radiation induced cellular reproductive death and chromosome aberrations. Radiat. Res., 76:573–586, 1978.

22. Terzaghi, M., and Little, J.B.: Repair of potentially lethal radiation damage in mammalian cells is associated with enhancement of malignant transformation. Nature, 253:548–549, 1975.

23. Little, J.B.: Factors influencing the repair of potentially lethal radiation damage in growth-inhibited human cells. Radiat. Res., 56:320–333, 1973.

Chapter 12

Responses of Tissues and Organs to Irradiation

12.1 INTRODUCTION

The effect of irradiation on tissues and organs is the production of morphologic and/or functional changes. None usually is observed after irradiation of most adult tissues or organs with low radiation doses—that is, doses ranging from a few rads to a few tens of rads. The reason is that tissue or organ changes occur as a result of *significant* degrees of lethal cell injury which, except in embryonic, fetal tissues, is unlikely at these doses. Very low doses may cause malignant transformation in cells of various tissues and organs, but this is a response of *cells* in a tissue, not of the tissue itself.

Radiation sensitivity (D_o) of most mammalian cells does not vary significantly, and with the exception of certain hemopoietic and lymphatic cells, neither do values of n. Since cell killing is the mechanism underlying tissue and organ responses, the relative uniformity of radiosensitivity and extrapolation numbers leads to the expectation that most tissues and organs will respond to various dose levels of radiation in about the same way. The contrary, however, is true; there is considerable difference among tissues and organs in response to irradiation.

12.2 PROLIFERATIVE VERSUS QUIESCENT CELLS

When tissues and organs are irradiated, proliferative cells principally are affected. As a rule, such cells are the stem cells of a tissue and are undifferentiated. Tissue function, however, is carried out by cells which are differentiated and have specialized in the performance of a particular function or group of functions. Differentiated, specialized cells, however, are not proliferative. At the end of their life span they die or are lost from the tissue without producing descendants. In some tissues these cells are not replaced when they die or are lost, and the number of functioning or specialized cells declines with time. This is the case with mammalian ova. At birth, female mammals have all the ova they will ever have. Through reproductive life, these mature and are shed from the ovary, and over the years, the number of remaining ova shrinks.

The same is presumed true of cells of adult mammalian central nervous system. Over the lifetime of an adult, some of these cells probably die, and the total decreases because there are no replacements. In most other tissues, however, as mature, functioning cells die, they are replaced. The number of functioning cells, therefore, remains more or less constant, because the supply is continually replenished to compensate for loss.

Replacements come from *stem* cells, undifferentiated cells which are proliferative. Because stem cells are undifferentiated and unspecialized, they cannot serve directly as replacements for differentiated, functional cells. First, they must differentiate and specialize to be able to perform the specific functions of cells they are to replace. As stem cells of certain tissues differentiate, they may also proliferate, so the process of differentiation in some tissues is accompanied by multiplication. Tissues whose mature, functional cells have short lives must replace the cells at a high rate. Examples include lymphatic tissue and gastrointestinal mucosa. Some tissues have functional cells with very long lives. These are replaced at very low rates (if they need to be replaced at all) because they are lost so slowly. Examples include central nervous system and muscle tissue. In the case of the central nervous system, the lives of mature cells are so long that there appear to be no proliferating stem cells at all. As for muscle, mature cells have very long lives, but replacement does occasionally occur.

The response of mammalian tissues to irradiation is controlled at least partly by the *length of life of functional cells* and, to a lesser degree, by innate sensitivity of cells in the tissue. Tissues with functional mature cells having short lives respond sooner after irradiation than those with long-lived functional cells. They have many proliferative cells, because the need to renew the supply of short-lived functional cells demands high levels of reproductive activity among stem cells. Mitotically static tissues respond minimally directly after irradiation; most of these cells are quiescent and, as such, relatively resistant to irradiation.

12.3 TISSUE RESPONSES

While irradiation *depopulates* the stem cells, that is, the mitotically active cells in a tissue, the *response* of tissues results from loss of and failure to replace functional cells. Functional cells, apparently unaffected by irradiation, live out their normal life span and die. Radiation depopulation of the stem cell compartment interrupts the flow of cells destined to become new functional cells, and the number of new functional cells being produced declines. A reduction in the total number of cells capable of performing tissue function results, and the performance of the tissue itself may be impaired.

12.4 TIME TO EXPRESSION OF TISSUE INJURY

The length of time that elapses between irradiation and expression of radiation injury, namely impairment of tissue function, depends on how long the *differentiation* process takes in a given tissue. Differentiation is the process by which stem cells become functioning cells, and if it takes a long time, it can be quite a while after irradiation before decline in tissue function is observed. In some tissues there are many steps in differentiation and a large number of cells differentiating at any given time. Differentiating cells are said to be *in transit* between the undifferentiated, stem-cell state and the final, differentiated, functional-cell state. In some tissues, hemopoietic tissues for example, there are many cells in transit at given times. Irradiation depopulates *stem* cells, but cells already in transit may complete specialization after irradiation. For a substantial period of time, new functional cells may continue to be produced, not from proliferating stem cells, but from cells already in transit. Eventually, however, the transit states become depleted as few new cells are fed into them from the depopulated stem-cell compartment, and the supply of functional cells declines. At that time, tissue function will decline and radiation injury (loss of tissue function) will be expressed.

Tissues differ in this regard. While in hemopoietic tissue, many cells are in transit and the transit time is long, in gastrointestinal mucosa the opposite seems to be true. Impairment of function may take several weeks to be observed as a consequence of damage to hemopoietic tissue but may be seen after only a few days as a result of damage to the gastrointestinal mucosa. The *time* required for expression of radiation injury of tissue is evidently determined by transit time. When it is long, much time may elapse before impairment of tissue function is observed. When it is short, impairment of tissue function may be observed soon after irradiation.

12.5 REPOPULATION

Though the number of cells in various tissue compartments (stem, transit and functional) may decline after irradiation, given time, each compartment may repopulate or be repopulated. The stem-cell compartment is the self-renewing component of tissues. In the normal course of events some cells produced in that compartment replace dead functional cells. However, if undifferentiated stem cells simply differentiated to become functional cells, the supply of stem cells would eventually decline. That compartment would become depleted, and the transit and functional compartments would follow suit. After irradiation, repopulation of all compartments may occur, beginning with the undifferentiated stem cells. Surviving stem cells multiply and those that can, produce a new clone of cells. These clones together repopulate the compartment, presumably eventually restoring it to the usual number of cells.

Not all surviving stem cells can produce viable clones. In some there may be chromosome damage which, either at the first or one of the subsequent mitotic divisions, causes them to die. Cells carrying such injuries may produce a few descendants, but

the chromosome injury in many cases causes the clone to die out.

The length of time needed for adequate repopulation of the stem-cell compartment depends on how many cells are lost from it as a result of irradiation and how long it takes for those cells to be lost. In part, this depends on the radiosensitivity of these cells. Certain hemopoietic and lymphatic cells are quite radiosensitive, and many of these will be killed by radiation doses that may not kill many other cell types (e.g., skin, gut). However, as mammalian tissues go, hemopoietic and lymphatic cells are exceptional; the sensitivity of most proliferating mammalian cells is roughly the same $(D_o = 150 \text{ rads})$. In most cases the number of cells lost from the stem-cell compartment immediately following irradiation will not depend so much on relative radiosensitivities as on the degree of proliferation in that compartment. Where radiosensitivities are equal, or approximately so, the number of cells killed by doses high enough to cause cell death will be controlled by the number proliferating.

12.6 MITOTIC DEATH

In addition to cells killed outright by radiation, there will be a number in which chromosomes will have been damaged. Broken chromosomes apparently do not noticeably affect cell function; however, when cells with damaged chromosomes proliferate, the damage to chromosomes may, either then or a few cell divisions later, result in cell death. The number of proliferating cells in a tissue also governs how many die from this effect.

Many cells in tissues with much proliferative activity die because damaged chromosomes prevent extended survival. Stem-cell compartments of tissues respond to loss of cells within the compartment by proliferation of surviving cells in the compartment. Stem-cell compartments that *lose* many cells as a result of irradiation with given doses are tissues in which there are already many proliferating cells. They suffer great cell loss precisely because they are active mitotically and must respond to this loss (repopulate) by proliferation of survivors. Among those survivors will be some with chromosome damage, and these will die when they proliferate. Mitotically active stem-cell compartments lose cells during irradiation, because proliferative cells are sensitive. They lose cells again when they repopulate, because chromosomally damaged cells then die. Ultimately the stem-cell compartment will repopulate and in the process will have been purged of cells with damaged chromosomes. Cells produced in the repopulated compartment, which then repopulate the other compartments, come from undamaged and/or repaired cells. In contrast to this are tissues in which the stem-cell compartment is less mitotically active. Fewer cells will be killed by given radiation doses than in tissue in which there is much mitotic activity, and cells with broken chromosomes may survive a long time.

Although functional cells may have long lives in such tissues, with the exceptions noted earlier, they do die and must eventually be replaced. Cells with injured chromosomes in the stem-cell compartment may at that time be called upon to divide to replace functional cells which die in the natural course of events. These stem cells, which have "harbored" radiation injury in the form of injured chromosomes, then die. The result is failure to replace lost functional cells *and* the loss of stem cells as well. *Repopulation* both of functional- and stem-cell compartments will be impaired since broken chromosomes, although "stored" a long time, eventually take their toll. Tissues having functional cells with long lives express radiation injury *late* after irradiation; tissues with rapid cell turnover express it *early* after irradiation.

12.7   ORGANS

Much of the foregoing discussion has dealt with radiation and tissues. Organs are also affected by irradiation; their responses are usually a composite of the responses of the tissues that make them up.

Organs are composed of a number of tissues. At the least there will be parenchymal (the tissue type that carries out organ function), connective, and vascular tissues. The response of organs to irradiation will be determined by the relative sensitivities of cells in the tissues comprising them and the rate of renewal of cells in these tissues. Hemopoietic and gastrointestinal tissues are rapidly renewed; there is a great deal of mitotic activity in them because of the continuing need to replace functional cells. Supporting stroma, the connective and vascular tissues which support hemopoietic and gastrointestinal organs, are more slowly renewed. Irradiation at *relatively* low doses damages active parenchymal cells and kills many. Since they renew rapidly, they quickly repopulate from cells which either were not damaged or have repaired their damage. Function of the organ, if compromised, is usually quickly restored. Later, however, another form of radiation damage may be manifest in these organs. Slowly renewing cells of supporting tissues eventually die and must be replaced. The *replacement* process may be impaired, because damaged cells die during the mitotic divisions basic to repopulation. Supporting tissue function is thus impaired. When vascular tissue is involved, changes in *organ* function due to hypoxia and, perhaps, parenchymal-cell malnutrition can result. Organs like the gastrointestinal tract, hemopoietic system, and skin may, therefore, respond twice to irradiation—early, when the number of parenchymal cells is reduced, and again later, when supporting tissue function is impaired.

The situation may differ in other organs. In those in which parenchymal tissue is very slowly renewed—or not renewed at all—such as the liver and kidney or in muscle and central nervous tissue, moderate to high radiation doses may fail to elicit even minimal early response. Late response may be the only one observed when, weeks or months after irradiation, supporting vascular and connective tissue function is impaired. These are the radiobiologic principles which many believe underlie early and late reactions of normal tissues following radiotherapy.

SUMMARY

1. Tissue and organ damage usually is expressed as impairment of function.
2. Radiation generally affects proliferating stem cells in tissues. Quiescent functional cells rarely are affected.
3. The number of proliferating cells in tissues usually is determined by the life span of functional cells; where the life span of functional cells is short, the need to replace them is great, and the tissue will be mitotically active. The reverse is true for tissues with long-lived functional cells.
4. In mitotically active tissues, the number of cells is sharply reduced immediately after irradiation; in static tissues, fewer cells are killed.
5. Depletion of mitotically active stem cells results eventually in reduction in number of quiescent functional cells, but the length of time required is dependent on the time needed for differentiation (transit time).
6. Repopulation of stem-cell compartments of mitotically active tissues takes place quickly and is accompanied by death of chromosomally injured cells (mitotic death).
7. Repopulation of stem-cell compartments of static tissues directly after irradiation is usually unnecessary, but cells with injured chromosomes remain in the tissue for a long time (no mitotic death).
8. Except for absolutely static tissues (central nervous system), functional cells must be replaced in all tissues. Stem cells with injured chromosomes die rather than replace lost functional cells, and the number of both stem and functional cells declines. In static tissues this occurs long after irradiation.
9. In mitotically active tissues, radiation injury is expressed soon after irradition; in static tissues, late after irradiation.
10. The response of organs to irradiation depends on the responses of the tissues composing them. If the parenchyma is mitotically active, it may respond early after irradiation. Stromal tissues are generally rather static and express injury late after irradiation. Organs with inactive parenchyma may show only the late, stromal response.

Chapter 13

Responses to Total-Body Radiation

13.1 INTRODUCTION

The response of organisms to total-body radiation depends on two main factors—whether or not radiation is given uniformly over the total body and the dose. When these two factors are controlled, several other factors—biologic variation, age, state of health, diurnal variation, dose rate, LET, and dose fractionation—can modify expected responses.

13.2 UNIFORMITY OR NONUNIFORMITY OF IRRADIATION

Nearly all that is known about the response of organisms to total-body radiation comes from studies using external radiation sources. Sources of radiation taken internally are usually deposited in only one or a few organs, irradiate only a few of its many tissues, and fail to produce a total-body radiation response.

Responses to nonuniform total-body irradiation from external sources are comprised of radiations' effects on all irradiated tissues, but the response of the organism is likely to be dominated by the responses of the most heavily irradiated tissues. Uniform irradiation produces total body effects dominated by the tissues most affected by given doses.

13.3 ACUTE OR LATE EFFECTS

Early (acute) adverse effects are those that result in clinical signs and symptoms within a few days or weeks after rapid, intense exposure to ionizing radiation. These effects, which occur in humans only after acute whole-body doses of at least 50 rads, increase in type and severity with increasing dose. They include gastrointestinal distress (nausea, vomiting, modification of gastric emptying time, diarrhea); hemorrhage; fluid imbalance; anemia; and infection. In addition there is often weight loss, loss of hair, and inhibition of gametogenesis. If dose is increased, a dose level will ultimately be reached that produces most or all of the foregoing and, in addition, kills some irradiated individuals, usually within a few weeks.

Late effects are those which are expressed months to years after irradiation. These effects are seen either after acute sublethal exposure (those which remain or develop after the early phase of tissue injury and repair) or after long-term or fractionated exposure. However, it is not always easy to sharply distinguish between early and late effects. The main late effects are life shortening, carcinogenesis, local damage to tissues (cataracts, sterility), effects on growth and development, and hereditary effects.

13.4 INFLUENCE OF DOSE ON MORTALITY (LD$_{50}$)

Dose versus mortality responses generally gave a sigmoid appearance (Figure 13.1). For mammals, most deaths have occurred within 30 days after acute total-body irradiation. Note that the upper and lower portions of the curve (Fig. 13.1) are poorly defined. The midpoint of the curve is used to characterize the dose-response relationship and is referred to as the *median lethal dose* which is ordinarily abbreviated as LD$_{50}$. Since it is a dose derived from a group of animals, it has relevance only to groups and not to individuals. Individual irradiated organisms may deviate markedly from the group as a whole, but the use of LD$_{50}$ tends to discount effects of individuals whose survival is markedly different from the group. Thus, LD$_{50}$ does not indicate that an individual exposed to this dose will necessarily live or die; it only indicates that within a large group, *half* of those exposed to the LD$_{50}$ are expected to die within a stated time, say 30 days (LD$_{50/30}$).

When expressed as the midline absorbed dose, large mammals (such as swine, burros, dogs, and probably humans) have LD$_{50}$ doses around 250 to 300 rads, while for small animals (rats, mice, rabbits), it is about double this value. Looking at Fig. 13.1, results of human exposure seem to anchor the lower end of the dose-mortality curve in the vicinity of 200 rads (where there is what amounts to a practical threshold). Using this value of the lower part of the curve, and assuming the slopes for large animals and for human beings to be about the same at higher doses, an estimated LD$_{50}$ of about 350 rads is obtained. Most of the

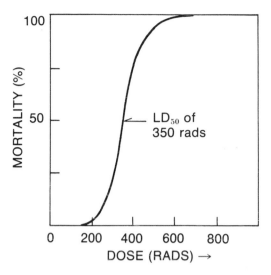

Fig. 13.1. A schematic presentation of the dose-mortality curve for humans in the absence of therapy for injury. The dose range between no deaths and 100% mortality is about 400 rads with an estimated LD$_{50}$ of about 350 rads.

scanty human data for doses greater than 200 rads are consistent with this value.

13.5 THE ACUTE RADIATION SYNDROMES

An hour or two after exposure of a major part of the body to a dose of radiation in excess of about 50 rads, transitory symptoms, called the *prodromal reaction*, begin to appear (Fig. 13.2). The time of onset and severity depend to some extent on individual susceptibility, but the principal influencing factor is radiation dose. The *approximate* doses eliciting signs and symptoms in 50% of those exposed are: anorexia, 120 rads; nausea, 170 rads; vomiting, 210 rads; and diarrhea, 240 rads.

The prodromal reaction is followed by the so-called *latent period*, a relatively asymptomatic period, the length of which depends upon the time required for development of sustained disturbances in organ function related to functional cell depletion. An exception is lethal brain damage which presumably is not due to func-

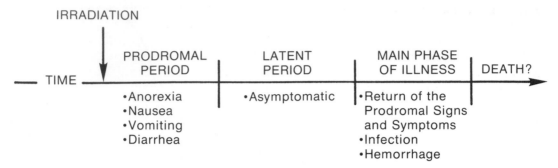

Fig. 13.2. The time course of the acute-radiation syndromes. At higher doses the latent period becomes progressively reduced and may not occur at very high doses. At the higher doses related to the CNS syndrome, the prodromal and main stages of illness will also have signs and symptoms related to CNS damage.

tional-cell depletion. As dose increases, the latent period becomes shorter and may not exist after very high doses.

The *main phase* of radiation-induced illness returns after the latent period and, with increasing dose, death becomes more likely as an outcome and occurs sooner after irradiation.

The acute effects fall into three rather distinct subgroups or syndromes (Table 13.1). The dose ranges given for these syndromes are, at best, estimates, since there is (fortunately) little human experience with exposure to high doses of radiation *alone*.

(Japanese exposed to the higher radiation doses from the atomic bombs were also subjected to thermal burns and blast injury).

The three subgroups or syndromes, arranged in order of increasing dose, are as follows: those in which the organ system primarily responsible for death is the *hematopoietic system;* those in which the organ systems primarily responsible for death are the *hematopoietic* and *gastrointestinal systems;* and those in which the organ system primarily responsible for death is the *cerebrovascular system.*

Table 13.1. *Acute radiation syndromes*[2]

Syndrome	Dose Range (Rads)	Duration of Latent Period	Pathology	Signs and Symptoms	Survival
Hematopoietic	200–700	1–3 weeks	Bone marrow aplasia Leukopenia Thrombocytopenia Anemia	Hemorrhage Infection	Survival possible Dose-related death within 2–6 weeks
Gastrointestinal	700–5000	1–4 days	Mitotic arrest in epithelium Denudation and ulceration of intestinal wall	Fever Diarrhea Disturbance of fluid and electrolyte balance	Survival highly improbable Death within 3–14 days
Central nervous system	5000–10,000	0–a few hours	Vasculitis Meningitis Encephalitis Edema Pyknosis of cerebellar granule cells	Apathy and drowsiness followed by tremor Convulsions Ataxia	Death within 2 days

13.6 HEMATOPOIETIC SYSTEM FAILURE

This response and its associated mortality, known also as *the bone marrow syndrome*, occurs in most mammals after doses of 200 to 1000 rads, (the range for humans is probably about 200 to 700 rads). *Mean survival time* of irradiated groups is dose responsive, ranging from four to six weeks at the lower end of the dose range to less than one week at its upper limit. There is a complex of signs and symptoms, but death occurs because the hematopoietic system, in mammals primarily confined to bone marrow, fails to produce mature functional cells for the circulating blood.

When mammals are totally irradiated, some bone marrow cells are killed outright (premitotic death), and others are delayed in division. Division delay is, of course, a common result of irradiation, and the time is probably used by cells to repair radiation damage. The proportion of bone marrow cells undergoing premitotic death increases with dose, leading to the supposition that such cells are so damaged they cannot live, let alone repair the damage. Also, as dose is increased, the length of mitotic inhibition increases, probably another reflection of increased damage.

Similar things doubtless occur in many other tissues of the body, but the effect in bone marrow stands out because it is the most radioresponsive *vital* system. It is radioresponsive partly because its cells are innately radiosensitive, but also because there are many cells in division.

Whether organisms die as a result of loss of function of bone marrow depends on whether recovery of damaged bone marrow cells occurs quickly enough for them to divide, differentiate, and replace functioning cells which are lost from irradiated individuals at about the same rate as from unirradiated individuals. The loss of bone marrow function ultimately means the loss of an organism's ability to combat infective elements and loss of the ability of blood to

clot, because the supply of new granulocytes and platelets is cut off. These factors are the cause of death as they lead to infection, nearly always from the organism's own intestinal flora, and the risk of hemorrhage.

At death all tissues have undergone degenerative changes, and some may be as severely affected as bone marrow. Among these are the gametogenetic systems, the lymph nodes, and possibly the hair follicles. The important difference is that none of these is vital; bone marrow is.

13.7 GASTROINTESTINAL SYNDROME

As dose increases above 700 to 1000 rads, more and more cells of the gastrointestinal tract (particularly the small intestine) become damaged by irradiation. From dose levels of approximately 700 to 1000 rads up to about 5000 rads, gastrointestinal injury is the *principal* cause of death. Those who receive a dose large enough that gastrointestinal damage results in death will have received far more than enough radiation to have resulted in hematopoietic death. Death from gut damage occurs, however, before the full effect of radiation on the blood-forming organs is expressed.

The mucosa of intestinal villi consist of four regions which house the organs' cell compartments. At the base of the villi are the crypts of Lieberkühn, the stem-cell compartment, a region of high mitotic activity. Somewhat higher along each villus is another region of high mitotic activity, the differentiating compartment. Near the villus tip are the functional absorptive cells which are produced in the crypts and have differentiated in the differentiating compartment. At the tip itself, spent functional cells are extruded from the villi into the intestinal lumen. Ideally, the number of cells lost from the tip equals the number produced in the crypt.

In most mammals, after doses in the range of 1000 rads, large numbers of crypt cells and cells of the differentiating compartment are either killed or mitotically delayed. Spent functional cells are nevertheless extruded from tips of villi at approximately the same rate as in unirradiated animals (actually the rate does slow somewhat in an apparent compensatory effort), but few new ones are produced in the crypts to replace them.

In time, the continuity of the intestinal lining is breached as spent cells leave and are not replaced. Eventually, these breaches become gaps of denuded intestine. Through these gaps, intestinal flora penetrate the organism, infecting it. The lack of mature granulocytes resulting from radiation's effect on bone marrow permits infective organisms to spread and multiply. Also, from these gaps leakage of blood occurs, and the short supply of platelets fails to stop it.

Finally, the loss of absorptive cells seriously impairs the capacity of the distal end of the small intestine to resorb bile salts. These pass into and irritate the large intestine which causes copious, watery diarrhea.

In most mammals, death occurs between three and four days after irradiation (in humans, although data are few, this time appears to be somewhat longer) as a direct result of infection, fluid and plasma loss, and salt imbalance, all consequences of severe diarrhea and vascular collapse. Mean survival time is evidently not dose dependent in the range of doses (about 1000 to 5000 rads) that produces death.

13.8 THE CEREBROVASCULAR OR CENTRAL NERVOUS SYSTEM SYNDROME

As dose is increased from 5000 to 10,000 rads (depending on the animal model tested), mean survival time again becomes dose dependent. This is a signal that death is no longer caused by loss of gastrointestinal function. The mechanisms responsible for death are not nearly as clear as they are at lower doses. The signs and symptoms comprising the syndrome point toward the central nervous system, but there is little gross or histologic evidence of what happens. In addition to diarrhea and vomiting, which probably result from effects on the digestive tract, signs include agitation alternating with apathy, disorientation, loss of balance, tetanic spasms, convulsions, prostration, and coma.

The time of onset depends on dose, but even during irradiation, specimens of some mammalian species display hyperactivity and irritability, occasionally alternating with spells of apathy.

Even during periods of relative quiet, convulsive seizures can be set off using minimal stimuli. As the syndrome progresses, more tremor, vomiting, watery diarrhea, and hysteria occur. Prostration and coma follow and then death occurs, usually within one to two days.

Most attribute the cause of death to brain edema which they believe accompanies very high irradiation doses. There is considerable evidence of infiltration of fluid, granulocytes, macrophages, and mononucleocytes into the meninges and brain. There is liquid infiltration into the choroid plexus, but the cause of these events is uncertain. Bacterial invasion seems unlikely because infiltration occurs so soon after irradiation. More likely, radiation damage to capillaries and possibly larger vessels, such as arterioles in the brain itself, causes the leakage. This fact prompts some to call this syndrome the cerebrovascular syndrome in recognition of the role of blood vessels.

Inside the skull there is little room for expansion, and fluid leakage is believed to cause pressure which in turn probably causes more tissue damage. The result is the cental nervous system syndrome and death.

Since so little is known for certain about this syndrome, explanations other than the foregoing can be devised to cover the observed facts. Some argue, for example, that extravasation of fluid and cells into the brain is the effect, not the cause, of the syndrome since much higher doses are required to produce the syndrome if the head alone is irradiated. Some feel high doses of radiation obliterate the blood-brain barrier, and numerous elements normally excluded from the brain can enter and affect it.

13.9 CARDIOVASCULAR SHOCK SYNDROME

Some consider that a fourth subgroup related to damage to the cardiovascular system also exists. In humans, after accidental overexposure, this phenomenon has been observed following high doses to the total body and about 5000 rads to the heart. In the documented cases, however, exposure has not been uniform as the upper body received the greater part.

Death, which occurs suddenly, has been attributed to cardiovascular shock. The workers who have identified the syndrome believe some human radiation fatalities, identified as results of central nervous system syndromes, may actually have been caused by cardiovascular shock.

13.10 TREATMENT OF ACUTE IRRADIATION

While damage to cells of all tissues of the body results in the total-body radiation syndrome, over a very wide dose range, death stems from a failure of two major organ systems, bone marrow and the gastrointestinal systems. The amount of damage produced in bone marrow and gastrointestinal systems depends on radiation dose. The greater the dose, the more damage there is, and the longer is the time required for the organ system to restore itself. Therapy of irradiated organisms usually involves replacement of cells of damaged organs or of the functions of damaged organs long enough for the ir-

radiated organism to become self-sufficient again.

13.11 POSSIBILITY OF SURVIVAL

Cronkite and Fliedner have classified radiation injury to humans into three clinical classes: survival improbable, survival possible, and survival probable.

Survival Improbable

Prompt and continuous nausea, vomiting, and diarrhea followed by virtual disappearance of neutrophils within five to six days with a steady decline of platelets to very low values by eight to ten days are very poor prognostic signs. The estimated dose producing this probably exceeds 500 rads and survival must be considered improbable with present therapy.

Survival Possible

These often are patients who have been exposed to relatively uniform whole-body irradiation in the range of 200 to 450 rads during radiation therapy. Nausea and vomiting subside within one to two days and are followed by a feeling of "well-being." These individuals later suffer primarily from the hematopoietic syndrome. They have been exposed to the dose range yielding very low to very high mortality (Fig. 13.1). In contrast to the previous group, their platelet counts diminish more slowly. Minimal values occur about four weeks after exposure, and there is slow recovery after this. An initial granulocytosis within the first two to four days continues to a plateau or rebound phase followed by a continued slow reduction over the first month. Slow recovery occurs after this.

Survival Probable

These individuals have received radiation doses of less than 150 to 200 rads (Fig. 13.1) and may have few or mild initial symptoms which disappear within a few hours. Unless marrow depression follows, there may be no further subjective effects of irradiation. Such individuals may, however, be susceptible to infection.

In the final analysis, treatment of the irradiated patient is basically the same as treatment of anyone with pancytopenia. Unlike many cases of drug-induced or idiopathic bone marrow aplasia, the aplastic marrow of the irradiated patient may be reversible, if the patient can be carried through the critical period.

13.12   TYPES OF TREATMENT

Therapy involves treatment of the signs and symptoms manifest in individual patients. With respect to the central nervous system syndrome produced by large amounts of radiation, there is little that can be done other than to give sedation and control symptoms.

Immediate Steps

A guiding principle in treatment of heavily irradiated individuals is to "do nothing without careful thought." For individuals who have not been so heavily irradiated that they are beyond all help, there are many hours, perhaps days, before therapy of any kind is required. A determination of radiation dose as well as a complete history and physical are needed. If neutron exposure is involved, there is some urgency in obtaining a blood sample so that dose can be estimated from neutron activation of sodium atoms in the blood. The lymphocyte count is valuable as an early criterion of radiation injury (Fig. 13.3A). If the lymphocyte count is 1200/mm³ or more 48 hours after exposure, it is unlikely that the individual has received a fatal dose; if the count is less than 300/mm³ at 48 hours post-irradiation, it is likely that injury will be fatal. Serial mitotic indices of bone marrow have also been suggested as an indicator of radiation exposure in the range of 50 to 200 rads. Although the number of

irradiated individuals that the data are based on is small, mitotic depression appears to be dose dependent in this dose range. A mitotic index approaching zero (the normal level at about noon is approximately 9 cells per 1000) at about four days post-exposure indicates a dose of 200 rads or more.

Treatment of Dehydration

At higher doses, watery, often bloody diarrhea (resulting in fluid loss and electrolyte imbalance believed to be responsible for death within a short time after irradiation) is often manifest. While gastrointestinal mucosa is regenerating, water loss and electrolyte imbalance must be treated to avert early death.

Treatment of Infection

Significant amounts of total-body radiation markedly reduce resistance to nearly any kind of bacterial infection. This increased susceptibility to infection is due in part to a breakdown in barrier function of various tissues (the gut in particular). There is also a correlation between time of onset of infection and depression of granulocyte count following radiation (Fig. 13.3C). Granulocytes are important in the body's defense against infection, and replacement therapy with granulocytes is a potentially effective therapy for uncontrollable infections in leukopenic patients.

When signs of infection develop, antibiotics in large doses are indicated. Different antibiotics should be tried until the fever is controlled. Sensitivity studies in culture should be undertaken, but one should not wait for the results since overwhelming infection can progress rapidly in the pancytopenic patient. Gamma globulin or immune serum may prove useful, although there is little information available on this point.

Changes in types of bacterial flora of the nares, throat, skin, and feces of men exposed to high levels of total-body radiation have been reported. Decreases were also

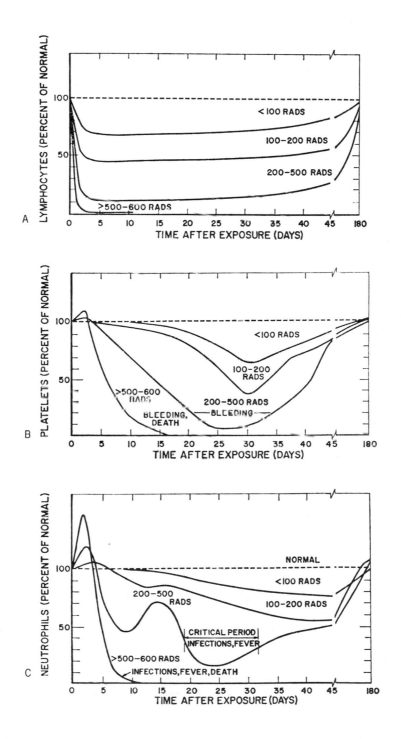

Fig. 13.3. The effect of acute whole-body irradiation on levels of: *A*, lymphocytes; *B*, platelets; and *C*, neutrophils. (From Langham, W.H.[6])

observed in agglutinin and bacteriocidal activity of their plasma.

Recent developments in plastic tent isolators combined with the use of antibiotics have made possible the maintenance of a relatively sterile environment. If the patient is placed in such an environment, it is worthwhile to get rid of as many bacteria as possible, as soon as possible. Oral antifungal antibiotics are useful in getting rid of bacteria of the intestinal tract. Caution should be exercised with this treatment since developing injury to the bowel may permit absorption of toxic antibiotics and sulfonamides which could do more damage than good.

Treatment of Bleeding

Bleeding has been demonstrated in acutely irradiated infection-free animals. In spite of the absence of infection, these animals die, and their deaths occur at a time when thrombocytes are markedly depleted (Fig. 13.3B). If fresh platelets are transfused, the animals do not die.

When repeated platelet transfusions are required to stop bleeding, there is a gradual decrease in the life of transfused cells (apparently an immunologic response of the host). Antibodies, although formed more slowly in totally irradiated than in unirradiated hosts, do agglutinate and destroy transfused cells. One must then decide *when* to begin platelet transfusions. If bleeding has already begun, it may be too late; if platelet transfusions are started too soon, the presence of platelet antibodies may reduce the effectiveness of transfused platelets when they are really needed. Observations of urine for microscopic hematuria have been suggested as a valuable index of bleeding tendency. If indications of bleeding occur, a single transfusion of platelets, equal in amount to that found in one-third of the patient's blood volume, has been recommended. After high doses of radiation, several massive transfusions at 3- to 5-day intervals may be indicated.

Bone Marrow Replacement

Because hemorrhage and increased susceptibility to infection are related to destruction of bone marrow, replacement of lost marrow cells would seem an effective therapy. This supposition is supported by animal experimentation which has shown that shielding a portion of bone marrow or spleen, or transfusion of unirradiated marrow, provides significant extension of life after irradiation. The increased probability that lethally irradiated, transfused organisms will survive irradiation depends on the *amount* of bone marrow transfused. The more transfused, the greater the chance of survival.

Not only does the amount of bone marrow transplanted have a great effect on the possibility of a "take," but survival of graft and host is also dependent on immunologic compatibility of host and donor—the more closely matched, the more successful and longer lasting the transplant.

The chance of protection by homologous transplant *increases* with increasing dose of radiation. At low doses, the chances of a "take" are reduced because immune responses of the host may quickly recover and quickly reject the graft. At high doses, immune response is deeply depressed or destroyed, allowing imcompatible bone marrow a chance of being therapeutic.

If homologous bone marrow "takes" in an irradiated individual, it is capable of supplying needed blood cells and allows the organism to survive acute effects of the irradiation. Later, however, the same organisms begin to die and, with time, die in increasing numbers. Death appears unconnected to radiation exposure. Rather, it appears to result from a reaction of donor cells against the host (the graft rejects the host). The response is called "secondary" or "runt" disease and is characterized by skin lesions, gastrointestinal dysfunction, atrophy of lymphoid tissue, and generalized "wasting" of tissues. The intensity of reaction and time of onset depend on the genetic relationship of host and donor. The more closely the two are related, the longer before secondary disease begins and the slower its progress. The reaction appears potentiated when small amounts of lymphocytic tissue are transfused with marrow.

Following any attempted homotransplantation in man, one desires a *temporary* take of bone marrow to save a life. It is hoped that there will be a slow recovery of the host's immunologic competence (which would reject the graft) and parallel recovery of the host's bone marrow. It is not clear how this result can be obtained. However, if the whole-body dose of radiation is proved without doubt to be in excess of 600 rads, there is almost no chance of spontaneous survival, and a bone marrow homotransplant should be considered.

13.13 LATE EFFECTS

Some individuals reveal one or more of the signs and symptoms of the total-body syndrome even in the range that does not produce an acute effect. Gastrointestinal distress and lowering of blood count are commonly observed during radiotherapy involving less than the total body. Effects on blood count reflect depression of cell production in marrow, but gastrointestinal distress is reported to occur even when the GI tract is not included in the irradiated field. The reasons are not known.

These effects are transient, and when they have passed, irradiated organisms go through a period of relative well-being. The period, variable in length and probably dose dependent, may end in one or more of the following late effects: carcinogenesis, impaired fertility, life shortening, and cataract induction.

13.14 CANCER

Exposure to radiation raises the incidence of cancer. This is true both in totally and in partially irradiated individuals. Considerable time may elapse between radiation exposure and the appearance of cancer, but *on the average*, irradiated individuals have a greater chance of getting cancer in the *irradiated area* than nonirradiated individuals (see Chap. 5).

13.15 IMPAIRED FERTILITY

Radiation may kill and inhibit division of precursor, premeiotic cells in the gonads. Because there is nearly always a supply of mature or maturing germ cells and because mature or maturing stages are resistant to doses up to 200 to 300 rads, directly after irradiation, fertility is normal. However, as mature cells are shed and maturing ones become functional and are shed, the number of new mature cells declines as a result of the loss of precursor cells that normally form them. Reduced numbers reduce the probability of fertilization, particularly in the male.

Since the doses in humans which produce permanent sterilization are high (about 500 to 600 rads in males and 300 to 400 rads in females) such an effect is unlikely after *whole-body* irradiation. (Such high doses are likely to be lethal before sterility becomes manifest.) In males, doses as low as 15 to 30 rads markedly reduce the sperm count at about eight weeks after exposure. The sperm count recovers slowly over the next few months. At doses above 100 to 150 rads, the sperm count begins to fall earlier, and after falling practically to zero may recover, but very slowly.

In males impaired fertility produces no associated loss of libido, but in females maturation of ova is accompanied by hormone production which stops or lessens with radiation-produced fertility impairment. Libido may diminish, and the accompaniments of menopause ("hot flashes," depression, absence of menstruation) may appear.

13.16 LIFE SHORTENING

Most animal experiments have shown that life span is shortened by radiation exposure. Studies have not yet unequivocally shown whether radiation's effect is primarily to accelerate the aging process or if the effect is related to induction of neoplasms. More recent studies of human and animal populations tend to indicate that the major component of radiation-induced life shortening comes from radiation-induced carcinogenesis. In the case of atomic bomb survivors, examination of causes of death and indicators of accelerated aging do not support the hypothesis that radiation accelerates aging per se. However, there is evidence that radiologists who entered the field before 1940 experienced a significant excess of early deaths when compared to the mortality of those in other medical specialties. There is increased death from cancer, but there is also increased *overall* earlier mortality, a finding so far unexplainable by radiation-produced pathology and consistent with the concept of accelerated aging.

13.17 CATARACTS

The ocular lens is enclosed in a capsule and covered with an epithelium anteriorly. Epithelial cells divide and migrate posteriorly to form lens fibers. Cell division in this epithelium continues throughout life, and this renewal system is subject to radiation damage at relatively low doses. Radiation damage to dividing cells results in abnormal fibers which migrate to the posterior portion of the lens producing a type of cataract characteristic of radiation damage. Since these cataracts can be distinguished from those from natural causes, the dose-effect relationship at lower doses may be determined more exactly.

The available data seem to indicate a threshold of about 200 rads of intense low-LET radiation (x rays, gamma rays) for the induction of detectable lens opacification in humans. The incidence of cataracts appears to increase linearly with doses above this threshold. For doses given at low dose rates or fractionated, the threshold is considerably higher, approximately 400 to 500 rads.

SUMMARY

1. As the dose of acute total-body radiation increases, signs and symptoms related to radiation damage occur. At doses of about 200 rads, some irradiated humans will die.
2. The early prodromal phase is followed by an asymptomatic period of variable length called the latent period. This is followed by the main stage of the illness and, if the dose is high enough, death.
3. Acute effects lead to death within a few weeks to a few minutes depending on the dose. Death occurs mainly from a failure of a critical organ system. In the human, in the dose range between 200 and 700 rads, death is from failure of the hematopoietic system; between 700 and 5000 rads, from failure of the gastrointestinal (and hematopoietic) system; and above 5000 to 7000 rads, from effects on the central nervous system.
4. Substitution or replacement of lost cells or functions can protect against the damaging effects of radiation.
5. Susceptibility to infection and bleeding can be reduced by granulocyte transfusion, by antibiotic therapy, fresh platelet transfusion, and isolation from pathogens.
6. Bone marrow transfusion will repopulate depleted marrow and protect against death, but graft-versus-host reaction and secondary disease are lethal, complicating factors.
7. Possible later effects of radiation in-

clude carcinogenesis, life shortening, impaired fertility, and cataracts.

REFERENCES

1. Andrews, G.A.: Radiation accidents and their management. Radiat. Res. (Suppl), *7*:390–397, 1967.
2. Upton, A.C.: Radiation Injury, Effects, Principles and Perspectives. Chicago, University of Chicago Press, 1969, pp. 1–31.
3. Pizzarello, D.J., and Witcofski, R.L.: Basic Radiation Biology. Philadelphia, Lea & Febiger, 1975, pp. 71–89.
4. Cronkite, E.P., and Fliedner, T.M.: The radiation syndromes. *In* Encyclopedia of Medical Radiology. Vol. II (3). New York, Springer-Verlag, 1972, pp. 299–339.
5. McLean, A.S.: Early adverse effects of radiation. Br. Med. Bull., *29*:69–73, 1973.
6. Report of the Space Radiation Study Panel of the Life Sciences Committee. Edited by W.H. Lang-ham. *In* Radiobiological Factors in Manned Space Flight. National Academy of Sciences, National Research Council Publication 1487, Washington, D.C., 1967, pp. 59–178.
7. Tsuya, A., et al.: Capillary microscopic observation on the superficial minute vessels of atomic bomb survivors, Hiroshima, 1972–1973. Radiat. Res., *72*:353–363, 1977.
8. Matanoski, G.M., et al.: The current mortality rates of radiologists and other physician specialists: specific causes of death. Am. J. Epidemiol., *101*:188–198, 1975.
9. Matanoski, G.M., et al.: The current mortality rates of radiologists and other physician specialists: specific causes of death. Am. J. Epidemiol., *101*:199–210, 1975.
10. Beebe, G.W., Kato, H., and Land, C.E.: Studies of the mortality of A-bomb survivors. 6. Mortality and radiation dose, 1950–1974. Radiat. Res., *75*:138–201, 1978.
11. Lushbaugh, C.C., and Casarett, G.W.: The effects of gonadal irradiation in clinical radiation therapy: A review. Cancer, *37*:1111–1120, 1976.
12. Ash, P.: The influence of radiation on fertility in man. Br. J. Radiol., *53*:271–278, 1980.

Chapter 14

Risks of Exposure to High Doses of Radiation

14.1 INTRODUCTION

In the context of this book, risks of exposure to high doses of radiation will be confined to risks resulting from situations in which a region of the body, generally small compared to the total body, is exposed to high doses of penetrating radiations, principally photons. Risks of exposure of the whole body or most of it to high radiation doses can become a medical problem after the fact, when accidentally irradiated persons require treatment. *Medically*, however, human beings are rarely given total-body radiation and are not commonly irradiated over very large portions of the body. Usually, relatively small regions are given high doses—divided into numerous fractions—to treat malignant disease. It is with the risks of such treatment that we are concerned here.

The risks encountered generally are known as reactions and/or complications of treatment. It is not within the scope of this book or the expertise of the authors to provide discussion of these complications. Rather, the *radiobiologic basis* or genesis of complications are summarized.

14.2 THE RISKS

Many of the risks have already been dealt with in earlier chapters, and the details will

not be repeated here. An overview or summary, however, may be useful.

Genetic Damage

If the gonads are in the irradiated field, severe impairment of fertility or even sterility can result. Cells involved in gametogenesis are quite radiosensitive and cumulative doses of several thousand rads can kill sufficient numbers of them to reduce the probability of fertilization nearly to zero.

Even when gonads are not in the treatment field, radiations are scattered to them and, depending on the dose absorbed this way, some degree of genetic damage is possible. The discussion of this phenomenon in Chapter 3 reveals that the degree of such damage expected in individual cases is *most* difficult to evaluate and predict. Mutation frequency increases resulting from the genetically significant dose (GSD) can be applied to large populations not to individuals and are no guide. Doubling dose (50 to 250 rads) can be used as a rough guide in predicting mutation frequency increase in individuals. If gonadal dose falls in that range, the result is expected to be a doubling of the spontaneous mutation frequency. Of course, doses above or below the range will probably result in some degree of change in frequency, but no precise figure can safely be estimated. Patients who have had gonadal irradiation during radiotherapy and who are considering becoming parents can be given such information, but how it helps them (if at all) in reaching decisions regarding potential parenthood is not clear.

As stated in Section 3.10, mutation frequency decreases with time after irradiation, reaching a low point at about two to three months post-irradiation. Patients may be advised to wait for that period before attempting conception. As an estimate of the possible effects in the first generation, a dose of 1 rad of parental gonadal irradiation is predicted to produce between 5 and 75 serious genetic injuries per million live-born, or 1 in about 30,000 per rad.

Oncogenesis

Radiation is a carcinogen, and it is reasonable to suppose that new malignancies may be induced during radiation treatment of an existing malignancy. Radiation-induced malignancies may occur in parts of the body not directly irradiated but to which radiations have been scattered, and they may also occur in normal tissues in a tumor's field of irradiation. Tissues outside the irradiation field receive relatively low doses and risk of induction can be estimated according to factors summarized in Chapter 5. Risk of cancers occurring in tissues *unrelated* to those in which a malignancy has already occurred is no different from the probability of occurrence of any primary malignancy, and it is reasonable to suppose that no special tendency toward radiation induction of cancers exists for unrelated tissues either.

New malignancies are in many cases more likely to occur in the irradiated field than outside the field, because tissues in the irradiated field receive fairly high doses. Exceptions to this generalization occur when the radiation dose in the field is high enough to kill most cells of coincidentally irradiated normal tissue, leaving so small a surviving population that tumorigenesis

is unlikely. Two factors govern whether this occurs—the dose itself and the kind of normal tissue. If coincidentally irradiated normal tissue is radiosensitive (e.g., bone marrow or lymphatic tissue), nearly complete sterilization is possible from doses used to treat malignancies. In less sensitive coincidentally irradiated tissues, such as the various soft tissues or bone, a sizable surviving population may remain containing significant numbers of transformed cells.

While radiation dose and tissue type govern the *number* of cells which may be transformed, *age* of the individual treated and length of life after treatment strongly influence the probability that a new neoplasm will actually appear. In Chapter 5 the latent period for tumor induction is discussed, and it is pointed out that, particularly in the case of nonhematogenous cancers, latent periods following radiation carcinogenesis are often years to decades long. When radiation therapy of children is successful, a long remaining life span can, in the natural course of events, be expected, and this may exceed the latent period for tumor development. Iatrogenic cancers may develop. On the other hand successful treatment of older persons may result in no increase in malignancies, because the remaining life span is shorter than the tumor's latent period. At any age, however, treatment which results in long survival presumably increases the chance that radiation-induced cancers will appear.

Studies performed so far indicate that malignancies induced in the irradiated field have not occurred in large numbers of cases. For example, although leukemia has been suspected of resulting from radiation therapy for Hodgkin's disease, a number of studies[1-3] have failed to produce convincing statistical relationships between radiation therapy for Hodgkin's lymphoma or multiple myeloma and late appearance of leukemia. The results of a study seeking a relationship between radiotherapy for ovarian carcinoma and later appearance of leukemia[4] were also negative, even though substantial fractions of bone marrow are exposed during treatment. Leukemia does seem to have occurred more frequently among persons treated with photons for Hodgkin's disease than among the population at large, but it appears that leukemia occurs in greater frequency among sufferers from Hodgkin's disease as a natural phenomenon,[2] and the observed higher incidence among those treated with radiation was not statistically significant. Even if one accepts (for argument's sake) *all* leukemias detected in these studies as having certainly been caused by previous radiation therapy for Hodgkin's disease or multiple myeloma or ovarian carcinoma, it appears that leukemia was induced in considerably less than 1% of treated cases.

Occasional reports of nonhematogenous cancers arising in excess in the irradiated field among persons treated for an earlier malignancy have appeared in the literature. Studies reporting statistically significant correlation[5,6] have involved malignancies arising 20 to 25 years after radiation therapy for cancer in childhood. Increases in *incidence* seem to have occurred, but because the natural incidence of such cancers is so low, even a large increase in incidence resulted in the fraction of those treated in which malignancies occurred of far less than 1%.

Nonlymphomatous malignant tumors have been reported, following irradiation for Hodgkin's disease, related to intensity of treatment.[7] Intensive radiotherapy seemed to increase incidence of such malignancies, but less intense treatment did not. Even here, however, the fraction of those treated in whom these malignancies occurred, was considerably less than 1%.

It appears that radiation therapy can increase risk of cancer in the treated volume, but that—for any *given* individual—the risk is not great. The benefit of successful treatment of a life-threatening malignancy certainly seems to outweigh this risk, for more than 99% of treated persons in the studies cited were cured without having a second malignancy.

These remarks apply *only* to situations in which radiation was the only treatment. Combined radio- and chemotherapy is probably more carcinogenic than either alone.

Developing Embryo and Fetus

A developing embryo and fetus may be exposed to radiation directly during radiation therapy if it is in the treatment field, or indirectly if radiations are scattered to it from a treatment field elsewhere in the maternal body. The fetal dose absorbed depends on just how close the fetus is to the treatment field and how large a dose is used.

During radiation therapy of pregnant women, cumulative fetal doses may easily exceed the 15 rads practical threshold given for production of congenital malformations (Chapter 6) and can, in fact, reach levels significantly above it. Consequently, radiation therapy of pregnant women may well increase significantly the hazard of production of malformations of development and/or growth retardation. However, convincing evidence of significant increases in occurrence of these phenomena is not easy to find, probably because the sample is so small. Relatively few pregnant women have been irradiated over the years. Nevertheless, the hazard the fetus carries of either subsequent malignancy or malformation or both seems almost certain to be increased and must be an important factor in weighing risks and benefits. For fetuses in the irradiated volume, risk of malformation is high, but for those to whom radiation is only scattered, it is much lower. For those in the field, the odds may even favor a severe malformation, but for those outside, while there is a greater than background probability of malformation, the odds are against it. Cumulative radiation doses of 200 rads to the fetus during organogenesis have nearly a 100% probability of producing an abnormality so large that it results in death at birth.

The presence of a pregnancy is a complicating factor in radiotherapy and, except where cumulative doses are expected to be less than about 15 rads, parents must be made aware of the potential risk to their child. If radiotherapy and chemotherapy are combined, the potential risk for the fetus is very much higher. The presence of a fetus may well be a strong, if not absolute, contraindication to the use of combined therapy because of the considerable hazard.

Loss of Tissue and/or Organ Function

Loss of normal tissue and/or organ function is probably the most important hazard that limits dose in radiotherapy. The responses of coincidentally irradiated normal tissues have been separated into two categories—acute and late. Acute responses, consisting of some degree of loss of function, occur soon after irradiation and are generally attributed to loss of parenchymal cells in rapidly renewing tissues (Chapter 12). Increasingly, late responses are noted. Loss of normal tissue and/or organ function occurs months or years after irradiation and may happen either in tissues which give no acute response or in those which have given and already recovered from an acute response. This latter hazard, once considered rare, is seen more often now, because of the greater success of radiotherapy. Greater extension of life has probably allowed more time for the complication to make itself manifest.

Acute effects are probably a less important dose-limiting complication than are late effects. They generally are transient, reasonably well understood, can be detected at their inception, and are fairly easily controlled. Late effects are evidently longer lasting, less predictable in occurrence, less well understood, and early indicators, if any, are not recognized. They can present great difficulties in clinical management.

There is dispute over the underlying mechanism of late effects. The predominant view is and has been that they result from radiation damage to the vascular system. Recently, however, a view has been advanced challenging this. Some believe[8] they result from late-manifested damage to parenchymal cells. The latter hypothesis holds that acute and late effects are due to the same basic mechanism, damage to parenchymal cells, but that *rate* of cell replacement in tissues causes the time of expression of such damage to be early in some tissues (acute responses) and quite late in others (late response).

Vascular Damage in Development of Late Effects

Studies in several animal models[9] indicate that circulation through small vessels in normal tissues, principally arterioles, becomes impaired following doses that produce tissue necrosis or atrophy. Moreover, time of impairment of circulation correlates well with the histologic appearance of atrophy or necrosis. Atrophy or necrosis, however, are not the late responses referred to here. These are transient phenomena and recovery from them occurs. Late *functional* impairment occurs after recovery from atrophy or necrosis. Impairment of circulation seems in some instances due to aggregates of cells, usually arteriolar or endothelial cells, occluding the lumina of these vessels, presumably resulting in hypoxia and tissue malnutrition. However, various other vascular injuries have also been proposed (e.g., narrowing of arterioles, hypertrophy of cells lining vessels, very small thrombi, changes in vessel wall permeability) to account for observed impairment of circulation. None, however, has been shown to be its sole cause. Nonetheless, adherents to the vascular hypothesis believe that vascular damage is the common denominator in production of late normal tissue changes.

Parenchymal and/or Stromal Cell Depletion

In opposition to the vascular hypothesis is a hypothesis which states that late effects result from loss, a long time post-irradiation, of either parenchymal or stromal cells of the irradiated tissue and not from impairment of vascular supply. The argument says that parenchymal or stromal cells in all tissues are injured during irradiation. In tissues with rapid proliferation rates (rapid renewal) lethal injuries that are expressed only when mitosis intervenes (for example, chromosomal damage) are expressed soon after irradiation. The high mitotic rate results in a high probability that given cells will either be recruited into cycle soon after irradiation or will be in cycle during irradiation. These tissues show an *acute* response to irradiation. In tissues with low proliferation rates (slowly renewing), the same injuries are expressed long after irradiation, because low mitotic rate results in a low probability of given cells being recruited into mitosis soon after irradiation or of being in mitosis during irradiation. Adherents to this hypothesis say that acute and late effects are manifestations of the same phenomenon. The *time* of expression is determined by proliferation rate in tissues.

Some arguments advanced to support the hypothesis are as follows. (1) Late effects tend to occur most often (if not exclusively) in slowly proliferating tissues such as nervous tissue, heart, connective tissue, lung, cartilage, bone, muscle, and liver. They are rarely observed in rapidly renewing tissues. (2) If vascular lesions were responsible for late effects, they would be expected to occur at about the same time post-irradiation in all tissues after the same dose. Vascular tissue is, after all, vascular tissue irrespective of the organ in which it is found, and it is reasonable to expect similar doses to produce similar injuries within similar time frames. In fact, late effects occur at various times after irradiation. For example, in lung they occur within six months of irradiation; in kidney, more than a year after irradiation; in the eye, between one and five years following irradiation. These latent periods are believed to correlate better with cell proliferation rates in the various tissues than with vascular injury. (3) There is a very broad range of doses required to produce late normal tissue injury, something the vascular hypothesis does not expect. The total dose required varies somewhat as a function of the size of each fraction and interval between fractions. It requires 200-rad daily fractions to produce late kidney effects after approximately 2000 rads total dose, about 5000 rads total dose to produce late effects in brain, and over 6000 rads total dose to produce late effects in prostate (Table 14.1). (4) In certain organs different tissue types are supplied by the same arterioles, yet after irradiation only one of the tissues shows a late response. If damage to arterioles were the mechanism underlying late responses, it is difficult to understand how one and not both tissues fed by the same arterioles would demonstrate an effect.

At present the cause or causes of late normal tissue effects is not known, but it is important to discover and understand them. It is essential that it be possible to predict their occurrence and either to prevent or treat them in order to deal with this dose-limiting complication in radiotherapy.

Residual Injury

When some irradiated tissues are re-irradiated months or even years after irradiation, it appears that these tissues have a lower tolerance than identical tissues that have never been irradiated. Such observations suggest that residual injury persists in these tissues and that additional irradiation adds to this injury, producing reactions at dose levels too low to have produced them in unirradiated tissue. Residual injury is evidently not the same as the sublethal damage that produces shoul-

Table 14.1. *Normal tissue tolerance*

Estimated Doses for 5% and 50% Incidences of Injury in Type 1 organs*

Type I Organs	Injury	$TD_{5/5}$† (rads)	$TD_{50/5}$‡ (rads)	Whole or Partial Organ (field-size or length)
Bone marrow	Aplasia, pancytopenia	250	450	whole
		3000	4000	segmental
Liver	Acute and chronic hepatitis	2500	4000	whole
		1500	2000	whole strip
Stomach	Ulcer, perforation, hemorrhage	4500	5500	100 cm²
Intestine	Ulcer, perforation, hemorrhage	4500	5500	400 cm²
		5000	6500	100 cm²
Brain	Infarction, necrosis	6000	7000	whole
Spinal Cord	Infarction, necrosis	4500	5500	10 cm
Heart	Pericarditis and pancarditis	4500	5500	60%
		7000	8000	25%
Lung	Acute and chronic pneumonitis	3000	3500	100 cm²
		1500	2500	whole
Kidney	Acute and chronic nephrosclerosis	1500	2000	whole (strip)
		2000	2500	whole

*Dose given assuming 200 rads/fraction, 5 fractions/week
†Dose for 5% injury in 5 years
‡Dose for 50% injury in 5 years
From Int. J. Rad. Oncol. Biol. Phys. Toxicology Working Group, 5:685, 1979.

ders on survival curves, for that type of sublethal damage is repaired within a few hours of irradiation. Neither can it be unrepaired potentially lethal damage (PLD) since that is repaired in even shorter times than sublethal damage. Residual injury does seem to be lost with time, however, since less of it appears to persist in irradiated tissues as the interval between irradiation and re-irradiation lengthens.

One type of residual injury that has been shown is damage to chromosomes. In a variety of animal and human tissues radiation-produced chromosome abnormalities have been demonstrated many months to years after irradiation.[10-12] It is possible that abnormal chromosomes produced by irradiation, remain in nondividing cells of various tissues. If these tissues are irradiated further at a later time, nondividing cells may be recruited into mitosis to repopulate cell compartments depleted by re-irradiation. Their damaged chromosomes may then make successful proliferation impossible, and these cells die (Chapter 4), causing tissue response at lower doses than

would have occurred after irradiating previously untreated tissues in which very few cells would contain broken chromosomes.

If this hypothesis is correct, repair or loss of residual injury should occur most quickly in tissues with rapid proliferative rates, because they will quickly purge themselves of cells with aberrant chromosomes after irradiation. Their rapid proliferative rate will mean that there is a high likelihood that *any* cell, with or without abnormal chromosomes, will be recruited into division soon after irradiation and many of those with damaged chromosomes will undergo mitotic death. That form of residual injury will be rapidly cleared.

On the other hand, tissues with low proliferative rates should retain this latent lethal injury for longer periods, but over time, the amount should slowly decrease as cells are slowly recruited into division. Experimental evidence partly supports these predictions. Early skin reactions are due primarily to damage to basal cells which proliferate rapidly after irradiation. This tissue is rapidly restored to near normal after irradiation, demonstrating little residual injury.[13] Hematopoietic tissues reveal little long-term injury[14–16] as does human skin.[17] Experiments involving slowly proliferating tissues, however, have not shown that residual damage is always repaired, however slowly, after irradiation. For example, no repair of residual damage appears to occur in the spinal cord of rats, and experiments involving deformity of mice have not always confirmed the prediction of the hypothesis.

The occurrence of slowly repaired residual injury is, nevertheless, an important hazard of high-dose irradiation in instances where re-irradiation is necessary or desirable.

SUMMARY

1. The delivery of high doses of ionizing radiation to small parts of the body, such as occurs during radiotherapy, presents certain hazards.
2. Radiations absorbed in gonads may increase mutation frequency. Gonads in the field of irradiation receive high doses, but those outside the field receive only scattered radiations.
3. Malignancies may be produced in normal tissues either within the treatment volume or outside it. Second malignancies occurring in the treatment field have occasionally been reported. The risk for individuals is small, probably less than 1%.
4. If fetus is irradiated during maternal exposure for treatment, fetal abnormalities may result. However, data are scanty due to the rarity of cases and firm conclusions are difficult to reach.
5. Loss of tissue and/or organ function may occur soon after or even during irradiation (acute response) or many months to years later (late response). Late response is a potentially serious complication of irradiation in long-

term survivors of treatment. There is no agreement as to its cause.
6. Residual injury may persist in irradiated tissue for a long time after irradiation. In rapidly renewing tissues the degree of injury declines rapidly after irradiation, but in slowly renewing tissues it drops less rapidly and persists a long time. In some tissues no loss of residual injury is detected. One form of residual injury may be chromosomal damage, but other forms may exist.

REFERENCES

1. Rosner, F., and Grunwald, H.W.: Hodgkin's disease and acute leukemia—Report of eight cases and review of the literature. Am. J. Med., *58*:339–353, 1975.
2. Rosner, F., and Grunwald, H.W.: Multiple myeloma terminating in acute leukemia—Report of twelve cases and review of the literature. Am. J. Med., *57*:927–939, 1974.
3. Coleman, C.N., et al.: Hematologic neoplasia in patients treated for Hodgkin's disease. N. Engl. J. Med., *297*:1249–1252, 1977.
4. Reimer, R.R., et al.: Acute leukemia after alkylating-agent therapy of ovarian cancer. N. Engl. J. Med., *297*:177–181, 1977.
5. D'Angio, G.J., et al.: Decreased risk of radiation associated second malignant neoplasms in actinomycin-D treated patients. Cancer, *37*:1177–1185, 1976.
6. Li, F.P., Cassady, J.R., and Jaffee, N.: Risk of second tumors in survivors of childhood cancer. Cancer, *35*:1230–1235, 1975.
7. Arseneau, J.C., et al.: Non-lymphomatous malignant tumors in association with intensive therapy. N. Engl. J. Med., *287*:1119–1122, 1972.
8. Withers, H.R., Peters, L.J., and Kogelnik, H.D.: The pathobiology of late effects of irradiation. *In* Radiation Biology in Cancer Research. Edited by R.E. Meyn and H.R. Withers. New York, Raven Press, 1980, pp. 439–448.
9. Hopewell, J.W.: The importance of vascular damage in the development of late radiation effects in normal tissues. *In* Radiation Biology in Cancer Research. Edited by R.E. Meyn and H.R. Withers. New York, Raven Press, 1980, pp. 449–459.
10. Weinbien, K., Fitschen, W., and Cohen, M.: The unmasking of chromosomal damage by regeneration of latent irradiation effects in the rat liver. Br. J. Radiol., *33*:419–425, 1960.
11. Curtis, H.J.: Biological mechanisms of delayed radiation damage in mammals. Curr. Top. Radiat. Res., *3*:139–174, 1967.
12. Savage, J.R.K., and Bigger, T.R.L.: Aberration

distribution and chromosomally marked clones in x-irradiated skin. *In* Mutagen-Induced Chromosome Damage in Man. Edited by H.J. Evans and D.C. Lloyd. Edinburgh University Press, 1978.

13. Field, S.B., Morrisey, S., and Kutsutani, Y.: Effects of fractionated irradiation on mouse lung and a phenomenon of slow repair. Br. J. Radiol., 49:700–707, 1976.

14. Porteous, D.D., and Lajtha, L.G.: On stem cell recovery after irradiation. Br. J. Haematol., 12:177–188, 1966.

15. Ainsworth, E.J., and Leong, G.F.: Recovery from radiation injury in dogs as evaluted by the split-dose technique. Radiat. Res., 29:131–142, 1966.

16. Mawes, C., Howard, A., and Gray, L.H.: Induction of chromosome structural damage in Erlich ascites tumor cells. Mutat. Res., 3:79–89, 1966.

17. Hunter, R.D., and Stewart, J.G.: The tolerance to re-irradiation of heavily irradiated human skin. Br. J. Radiol., 50:573–575, 1977.

Chapter 15

Risks of Diagnostic Ultrasound

15.1 INTRODUCTION

Ultrasound, as used in medical diagnosis, is becoming increasingly popular. Nearly every year more diagnostic examinations are performed using it and there is no reason to think increased use will soon abate. Such widespread exposure of the human population, some before birth, has raised the question of whether there is risk attached to the diagnostic use of ultrasound that may need to be weighed against its benefit. Unfortunately, research in this area is in too early a stage to give even the most tentative answers. It seems clear that diagnostic ultrasound produces no acute or immediate reactions. Large numbers of people are exposed daily, none with any perceptible adverse effects. Nor do large numbers of effects of any sort appear to occur in the years following its use. Diagnostic ultrasound has been in use for at least two decades, but no one can point to sharply increased frequencies of any spontaneously occurring abnormalities or to any unique abnormality arising in those exposed, either in utero or in postnatal life.

On the other hand, possible small increases in incidence of adverse effects certainly have not been and cannot yet be ruled out. The same, of course, has been said of ionizing radiations. No one can be *sure* that the low doses associated with diagnostic exposure produce adverse effects, but the speculations resulting in risk estimates for diagnostic radiation exposures are based on firmer ground than similar speculations regarding ultrasound exposure. As an example, techniques available for measuring absorbed dose of ionizing radiations in tissue produce *relatively* accurate measurements. Nothing equivalent exists for measurement of *absorbed* dose of ultrasound. Presently, only exposure conditions can be described. The absorbed dose that is produced by various exposure conditions is unknown, and whether the same exposure conditions always result in the same absorbed dose in similar tissues is also unknown.

As another example, the cellular target(s) of ionizing radiation has been reasonably well identified; most investigators believe that DNA, or the chromosomes or DNA attached to nuclear membrane, is the target that, when critically damaged, results in detectable effects. But how low levels of ultrasound may damage cells (if, indeed, they do) is far from clear, for no targets comparable to those of ionizing radiations have been identified.

Ultrasound, in itself and in its interactions with matter, differs substantially from

154

ionizing radiations. Ionizing radiations are extremely energetic particles or electromagnetic waves, traveling at great velocities, interacting in cells by depositing very small amounts of energy through energy exchanges with orbital electrons. Ultrasound consists of mechanical vibrations which probably do not deposit energy at the atomic level but at levels of greater complexity.

Ultrasound frequencies are above the range of human hearing, the lower limit being defined at about 20,000 Hertz (Hz) (1 Hz = 1 cycle per second). Most *diagnostic* ultrasound machines employ frequencies between 1 and 15 megahertz (MHz) or 1 and 15 million cycles per second.

The initial interaction between ionizing radiations and matter are fairly independent of the structure of matter. Interactions with orbital electrons do not depend on cell or tissue type or architecture. These, with some well-known exceptions, are not important factors in determining tissue dose during passage of ionizing radiations through living material.

Ultrasound may be quite different in this respect. Since it is a series of mechanical vibrations, these may be reflected, wholly or in part, from the interfaces between various tissues and may lose significantly lesser or greater amounts of energy as they pass through tissues varying in density. Energy absorbed from a beam of ultrasound may vary greatly in given tissues, depending on how much is reflected into them when they bounce back from adjacent tissue. A particular tissue lying next to bone, for example, may be exposed to a lot of energy because bone is highly reflective. The same tissue lying adjacent to fat might be exposed to considerably less energy.

Finally, while interactions of ionizing radiations and matter result in the production of many excited states (ions and free radicals) and very little heat; ultrasound, particularly in ranges above the diagnostic range, produces quite a lot of heat. How-

ever, the amount of heat produced by ultrasound in the diagnostic range is not known, and it is uncertain whether the heat it does produce can yield significant biologic damage.

There is little question that ultrasound can produce tissue damage; therapeutic use of sound depends on this phenomenon. There remain many unanswered questions about whether sound in the diagnostic range is damaging. Though much work has been done and some effects demonstrated in model systems, evidence that sound is damaging, as it is used diagnostically, remains to be produced.[1]

15.2 POWER AND INTENSITY

An ultrasound-emitting device (a transducer) vibrates when electrically driven so as to transfer energy from the device to the adjacent medium. Acoustic power is equal to the amount of acoustic energy a device produces in a unit of time. Power is measured in watts (W), or more commonly in milliwatts (mW), and is a measure of the rate of flow of energy per second from the transducer to the medium. When vibration of the ultrasonic device is uninterrupted, the device is said to operate in the *continuous*-wave or CW mode. For most diagnostic ultrasound imaging devices, ultrasound is not emitted continuously but as a series of short pulses separated by periods of no emission. Such pulsed ultrasound is described by the "duty factor," the fraction of the total time sound is "on."

For the pulsed unit, since power varies with time, average power is determined by multiplying peak power by the duty factor. Thus, a device with a peak power of 10 W and a duty factor of 0.0004 (on 0.04% of the time) would have an average power of 0.004 or 4 mW.

Acoustic power is usually measured by one of two methods—either ultrasound is directed into a thermally insulated absorbing liquid where ultrasound energy is converted to heat resulting in a rise in temperature; or the ultrasound beam is directed at an absorbing rubber disc suspended from a balance in a water bath (the radiation force is measured by the displacement of the balance).

Just as with light, intensity is of prime importance. Sunlight directed on the skin will warm, but if concentrated on a small area of skin by a lens (increased intensity), it will burn. Intensity or "power density" is the *concentration* of power within an area, usually expressed as milliwatts per square centimeter (mW/cm²). The intensity of a sound beam is of prime importance in the consideration of biologic effects. The measurement of intensity is complicated by its variation with time with the pulsed mode, and its variability throughout the beam. When total power is divided by the area of the transducer face, the value obtained is the average power intensity, the so-called spatial average (SA) intensity. However, the average intensity may be much less than the highest value of intensity, the so-called spatial peak (SP).

Spatial *average*, temporal average (SATA) intensity is the quantity most easily measured and hence is most often quoted, and as the name implies, it is an "average" value of the intensity averaged over time. The spatial *peak*, temporal average (SPTA) intensity is the intensity at the position of its maximum value averaged over time. The SPTA intensity is most commonly used when discussing potential biologic effects.

15.3 ABSORPTION OF ULTRASOUND ENERGY

In order to understand the potential for biologic effects it is necessary to appreciate, even if in a simplified form, the nature of the process by which ultrasound energy is attenuated in living tissues.

Absorption

This process involves the irreversible transfer of energy to molecules of the me-

dium (mainly macromolecules in living tissues) most of it resulting ultimately in *heat*. For diagnostic ultrasound intensities, most energy is lost by this process.

Cavitation

Another attenuation process, cavitation, results from formation of gas bubbles in the medium. All normal liquids contain large numbers of submicroscopic gas bubbles which tend to grow under the action of mechanical vibration as gas from the surrounding liquid is transferred to the bubbles. Upon reaching a certain size in relation to the ultrasonic wave length, they behave as a resonant cavity and vibrate with large amplitudes. Such rapid oscillations may cause streaming patterns and also result in the production of free radicals. Cavitation appears to be a phenomenon of the liquid state, and its occurrence within organized tissues at the intensities used in *diagnostic* ultrasound appears to be highly unlikely.

Other or Direct

This type of interaction (nonthermal and noncavitational) is poorly understood, but experiments have demonstrated biologic effects of ultrasound under conditions where neither temperature rise nor cavitation were significant factors. Such a direct interaction is of special interest since it is unlikely that ultrasound intensities used in medical diagnosis would induce either appreciable temperature rise or cavitation.

15.4 IS DIAGNOSTIC ULTRASOUND SAFE?

One reply to this question would be the observation that, even with its wide use, there is no *known* instance of human injury caused by diagnostic ultrasound. This would seem to imply a high level of safety; yet, is it "absolutely safe?"

The only other source of information on

potential biologic effects comes from experiments in which ultrasound was administered to experimental animals. Several biologic effects have been reported in recent years. These include a change in the ability of cells to attach to solid surfaces,[2] reduction in mitotic index in regenerating rat liver,[3] reduction in growth rate in regenerating forelimbs of newts,[4,5] mortality and growth retardation in rat embryos and fruit flies,[6,7] reduced transplantability of a murine tumor,[9] effects on DNA and growth patterns of animal cells in vitro,[10,11] and sister chromated exchanges following in vitro exposure to diagnostic ultrasound.[10] However, some of these reports have been criticized based on inadequacies of dosimetry or difficulties in precise repetition,[8] and the matter remains controversial. The Biologic Effects Committee of the American Institute of Ultrasound in Medicine has reviewed many of these reports and issued the following statement:

STATEMENT ON MAMMALIAN IN VIVO
ULTRASONIC BIOLOGICAL EFFECTS
August 1976: Revised October 1978
In the low megahertz frequency range there have been (as of this date) no independently confirmed significant biological effects in mammalian tissues exposed to intensities* below 100 mW/cm². Furthermore, for ultrasonic exposure times† less than 500 seconds and greater than 1 second, such effects have not been demonstrated even at higher intensities, when the product of intensity* and exposure time† is less than 50 joules/cm².

This statement describes SPTA intensity (100 mW/cm²) and time limits below which "independently confirmed significant biological effects" have not been observed. The intensity quoted, 100 mW/cm², must be considered in relation to the following:

1. Data reported for commercial pulse-echo diagnostic ultrasound units now in use show SPTA intensities in the range of 0.1 to 200 mW/cm².

2. It is derived from experimental studies and may be reduced as a result of further studies.

3. Clinical ultrasound studies usually involve scanning while experiments usually involve a stationary beam. The organs of a patient are usually farther from the sound source, and a smaller fraction of the organ is covered (because the experimental animal is normally much smaller). These conditions may provide some unknown safety factors.

4. The SPTA intensity limit is not a level which if exceeded will be unsafe, and below which one should feel completely safe with unlimited use. At this point it is simply a "guide."

Since we cannot now say that the medical use of ultrasound is "absolutely safe" we are left, as we are in many medical tests, with minimizing potential risks as compared to potential benefits. Generally this means the use of minimum exposure (intensity and time). When buying equipment, if two instruments are comparable in other respects, choose the one with the lowest intensity.

Finally ultrasound should *only be used when a benefit is expected*. This is particularly true during pregnancy because of the recognized susceptibility of the embryo to a wide variety of insults. For example, the use of Doppler ultrasound to obtain external tracing of heart rates during delivery involves use of a lengthy-exposure procedure in many low-risk pregnancies. Such use of ultrasound is difficult to justify since there are other ways to follow fetal heart rate with no risk.

SUMMARY

1. Diagnostic ultrasound *appears* to be relatively safe when compared to other imaging techniques such as x-ray studies. Safety, however, implies proof and there is no conclusive proof

*Spatial peak, temporal average as measured in a free field in water.
†Total time; this includes off-time as well as on-time for a repeated-pulse regime.

that it is safe, other than the lack of reports that ultrasound has harmed humans.

2. Ultrasound is not a toy. Specific medical benefits must be expected from its use.

REFERENCES

1. Wells, P.N.T., editor: Ultrasonics in Medical Diagnosis. 2nd Edition. New York, Churchill Livingstone, 1977.
2. Siegel, E., et al.: Cellular attachment as a sensitive indicator of the effects of diagnostic ultrasound exposure on cultured human cells. Radiology, 133:175, 1979.
3. Kremkau, F.E., and Witcofski, R.L.: Mitotic reduction in rat liver exposed to ultrasound. J. Clin. Ultrasound, 2:123–126, 1974.
4. Miller, M.W., et al.: Absence of mitotic reduction in regenerating rat livers exposed to ultrasound. J. Clin. Ultrasound, 4:169–172, 1976.
5. Wolsky, A., and Pizzarello, D.J.: Morphostatic effect of ultrasound on limb regeneration in the newt, *Triturus (Diemictylus) viridescens*. Am. Zoologist, 13:1350, 1973.
6. Pizzarello, D.J., Wolsky, A., Becker, M.H., and Keegan, A.F.: A new approach to testing the effect of ultrasound on tissue growth and differentiation. Oncology, 31:226–232, 1975.
7. Pizzarello, D.J., et al.: Effect of pulsed, low-power ultrasound on growing tissues. I. Developing mammalian and insect tissue. Exp. Cell Biol., 46:179–191, 1978.
8. Child, S.Z., Carstensen, E.L., and Davis, M.D.: A test for "miniature flies" following exposure of *Drosophila melanogaster* larvae to diagnostic levels of ultrasound. Exp. Cell Biol., 48:461–466, 1980.
9. Pizzarello, D.J., Vivino, A., Newall, J., and Wolsky, A.: Effect of pulsed, low-power ultrasound on growing tissues. II. Malignant tissue. Exp. Cell Biol., 6:240–245, 1978.
10. Liebeskind, D., et al.: Effects on the DNA and growth patterns of animal cells. Radiology, 131:177–184, 1979.
11. Liebeskind, D., Bases, R., Eleguin, F., and Koenigsberg, M.: Sister chromatid exchanges in human lymphocytes after exposure to diagnostic ultrasound. Science, 205:1273–1275, 1979

Index

Page numbers in *italics* refer to illustrations; page numbers followed by "t" refer to tables.

Abdominal cancers, radiotherapy of, 111
Abortion, radiation-induced, 40
  therapeutic, after radiation exposure, 72–73, 79
Absorbed dose, 3
Accelerators, radiations produced by, 11t
Acute radiation syndromes, 135–143, 149–150. See also specific
    syndromes
  latent period for development, 135–136, *136*
  survival possibilities of, 136t, 139–140
  treatment of, 139–143
Adenocarcinomas, 53
Adenoids, enlarged, x-irradiation therapy for, 57
Adenomas, radiation-induced, 57
Age. See also *Embryo; Fetus*
  genetically significant dose and, 20–21
  influence on radiation response, 2, 36, 56–57
    adults vs. children, 6
    breast cancer and, 79–81
    carcinogenesis and, 53, 147–148
    131I treatments and, 85
    pregnancy and, 71–72
ALARA concept, 91
Alleles, 17
Alpha particles, LET value of, 13t
  properties of, 11t
Anemia, radiation-induced, 136t
Anesthetics, radioprotection by, 109
Ankylosing spondylitis, radium treatments and, 59
  x-ray treatments and, 55t, 56
Anorexia, radiation-induced, 135
Apathy, radiation-induced, 136t, 138
Aplasia, radiation-induced, dose required, 151t
Arterioles, radiosensitivity of, 149
Ataxia, radiation-induced, 136t
Atomic bomb survivors, 54–56
  fetal irradiation and, 58, 78
  Life Span Study, 55
  radiation-induced cancers in, 53, 55t
    breast, 80t
    spontaneous cancers vs., 46
  radiation-induced chromosome aberrations in, 37, 37t
  radiation-induced congenital malformations in, 68
  radiation-induced point mutations in, 25–26
Atomic numbers, 8–9
Atomic reactors, radiation produced by, 11t
Atomic structure, 8–9

Background radiation, natural, cell killing, 117
  mutation frequency and, 16
Barium enema, genetically significant dose and, 21t
  radiation dose from, 5t
    embryonic dose of, 70t

BCDDP, 80–81
BEIR report. See *Biological Effects of Ionizing Radiation report*
Beta particles, properties, 11t
Betatron, radiation produced by, 11t
Biological Effects of Ionizing Radiation (BEIR) report, doubling
    dose, 24
  effect of fetal irradiation, 58
  low-dose radiation, 51–52
    lifetime cancer risk, 54t
  radiation-induced mutations, 27, 28t
Bladder, radiation dose from 99mTc-DTPA, 7t
  sensitivity to radiation carcinogenesis, 52–53, 52t
  risk in particular populations, 55t
Blood, radiation dose from 99mTc-albumin, 7t
Bloom's syndrome, 44
Bone, radiation dose from 99mTc-polyphosphate, 7t
  sensitivity to radiation carcinogenesis, 52–53, 52t
  x ray absorption by, 3
Bone cancer, radiation-induced, latent period of, 43, 43t
  radium dial painters and, 61
  radium treatments and, 59
  risk in particular populations, 55t
Bone marrow, mitotic index of, 140
  radiosensitivity of, 137
  radiotolerance of, 151t
  replacement of, 142–143
Bone marrow aplasia, radiation-induced, 136t
Bone marrow dose, 2, 4
  annual, in United States population, 77
  for specific radiographic examinations, 5t, 77
  131I treatments and, 85–86
  32P treatments and, 86
Bone marrow syndrome. See *Hematopoietic system failure*
Bone tuberculosis, radium treatments and, 59
Bragg peak, 13, *13*
  radiotherapy and, 110
Brain, radiosensitivity of, 138–139, 150
  radiotolerance of, 151t
Brain edema, radiation-induced, 138
Breast, sensitivity to radiation carcinogenesis, 52–53, 52t
Breast cancer, radiation-induced. See also *Mammography;
    Mastitis*
  age and, 56–57, 79
  atomic bomb survivors and, 56, 80t
  chest fluoroscopy and, 80t
  dose-response relationship, 79
  131I treatments and, 60, 85–86
  latent period, 43, 43t, 79
  mammography and, 79–81
  mastitis treatment and, 80t
  risk in particular populations, 55t
  spontaneous, 79–80
Breast Cancer Detection Demonstration Project (BCDDP), 80–81

Cancers, hypoxic cells in, 106–108, *108*
    radiotherapy and, 107–109
        hyperbaric oxygenation and, 109
    thermotolerance of, 115–116
    radiation-induced, 42–64, 147–148. See also names of specific
        cancers
        atomic bomb survivors and, 46
        detection of, 45–46
        dose-effect relationship of, 46–52, *48, 49,* 147
        in utero irradiation and, 76
            by x rays, 55t, 58
        latent period, 43, 43t, 147–148
            immune suppression and, 45
        mechanism of carcinogenesis, 43–45
        nuclear medicine and, 83–84
        occupational radiation exposure and, 91–92
        organ sensitivity to, 52–53, 52t
        origin of, multi-event, 44–45
        risk, from radiation therapy, 76, 147
            from radiographic examinations, 78
            in particular populations, 55t
            lifetime risk estimates, 52t
            measurement of, 52
            total lifetime, 53–54
        total-body radiation and, 143
    spontaneous, 42–46
Carcinogenesis. See *Cancers, radiation-induced*
Carcinogens, 43–45
    in vitro transformations by, 126–127
Cardiovascular shock syndrome, 139
Cataracts, radiation-induced, 92–93
    total-body radiation and, 134–135, 144
Cavitation, ultrasound and, 156
Cell life cycle, cohort cells and, 123
    density inhibition and, 126–127
    mitosis, 34–40, 122, *122*
    radiosensitivity and, 121–123, *122*
    thermotolerance and, 115
Cell-survival curves. See *Dose-survival relationships*
Central nervous system, radiation response of, 132
    radiation-induced congenital malformations in, 68
Central nervous system syndrome, 134, 136t
    pathology of, 136t, 138–139
    treatment of, 140
Centromeres, 32, *34*
    terminal, 38–40
Cerebrovascular syndrome. See *Central nervous system syndrome*
Cervical cancer, radiation-induced, 57
Chest fluoroscopy, 55t, 56
    breast cancer induction and, 80t
Cholangiosarcomas, induction by thorium, 59
Chromium, radioactive (^{51}Cr), radiation dose from, 7t
Chromosomal theory of malignant tumors, 44
Chromosome(s), 17. See also *Meiosis; Mitosis*
    arms of, 32, 38–40
    centromeres of, 32, *34*
    fragments, 33, *34*
    gene order on, 31–32
    in malignant cells, 43–45
    lagging, 38–40
    morphology of, 31–36
    radiation damage and, 36–40, 71, 125, 131–132, 151
    ultrasound damage and, 157
Chromosome aberrations, 31–40, 151
    in malignant cells, 39–40, 43–45
Cohort cells, 123
Coma, radiation-induced, 138
Congenital malformations, radiation-induced, 66–68, *66,* 76
    minimum fetal dose for, 67t, 72–73
    radiographic examinations and, 78–79
Connective tissue, radiosensitivity of, 132
Convulsions, radiation-induced, 136t, 138
Cooperative Thyrotoxicosis Study, 60
Corpuscular radiations, 10
Cosmic radiation, mutation frequency and, 16
^{51}Cr. See *Chromium, radioactive*
Critical organ dose, from radiopharmaceuticals, 6, 7t
Cyclotron, radiation produced by, 11t
    radiation-induced cataracts in workers, 92
Cystamine, radioprotection by, 114
Cysteine, radioprotection by, 114

DE. See *Dose Equivalent Unit*

Developmental defects, radiation-induced, 40
Diarrhea, radiation-induced, 135, 136t, 138
    treatment of, 140
Diathermy, 115
Dicentric chromosomes, 34–35
Differentiated cells, from stem cells, 129–131
    radiation response of, 129–133
    repopulation after irradiation, 131–132
Diffuse toxic goiter, 6
DNA, radiation damage to, 101, 125–127
Dose Equivalent Unit (DE), 3
Dose rate, 116–118, *117*
    chromosome aberrations and, 36–38
Dose-survival relationships, 120–121. See also *Post-irradiation
        conditions*
    calculation of radiosensitivity from, 99–100
    dose rate and, *117*
    exponential, 94–99, *95, 96, 98*
    extrapolation number of, 98–99
    in vivo measurements of, 124–125, *126*
    LET and, 94–99, 103–104
    oxygen effect and, 104–111, *105, 106*
    Q theory of, 101
    sigmoid, 94–99, *95, 96, 98*
        quasi-threshold radiation dose, 100, *100*
    single hit killing, 94, 96–99
    sublethal radiation damage, 95–99
    target theory of, 97–101
Doubling dose, definition of, 23–24
    gonadal dose vs., 27, 146
Down's syndrome, 40

Edema, radiation-induced, 136t
Electromagnetic radiations, 10
Electrons, 8, 11t
    LET value, 13t
    pairing of, 9
    valence electrons, 9
Embryo, radiation-induced congenital malformations in, 66–68,
        66, 67t, 148–149
    radiation-induced death of, 66–67, *66,* 76–79
        minimum lethal dose, 67t
    radiation-induced growth retardation of, 78–79
    radiosensitivity of, during organogenesis, 66–67, *66,* 69t,
        71–72, 148–149
        of differentiating cells, 65
        of gonads, 69
    therapeutic abortions of, after radiation, 72–73
    ultrasound exposure of, 157
Embryonic dose, from maternal ^{131}I, 84
    from maternal x-ray examinations, 70t
Encephalitis, radiation-induced, 136t
Enhancement ratios for radiosensitizers, 112
Ethylphosphorathionic acid, radioprotection by, 114
Extrapolation number, hits for cell killing, 98–99
        oxygen effect and, *106*

Fallout, radioactive, ^{131}I, 60
    Marshall Islanders and, 56
    Project Smoky, 51
    yearly individual radiation exposure to, 1t
Fanconi's syndrome, 44
Fertility. See also *Sterility, radiation-induced*
    maximum permissible dose and, 93
    radiation-induced impairment of, 25, 27, 143, 146
Fetal dose, 2, 55t, 58, 148
    from maternal ^{131}I, 70–71, 84
    maximum permissible dose and, 91
    radiographic examinations and, 70t, 78
Fetus, radiation-induced death of, 66–67
    radiosensitivity of, 53, 66–67, *66,* 69t, 148–149
        to ^{131}I, 70–73
    therapeutic abortion of, after radiation, 72–73
Fever, radiation-induced, 136t
    treatment, 140
Flagyl. See *Metronidazole*
Fluoroscopy. See *Chest fluoroscopy*
Free radicals, from water, 10
    radiation damage by, 10
        oxygen effect and, 106

radioprotectors and, 114
radiation-produced, 9–10
radiosensitizers and, 111–112
ultrasound-produced, 156

G1 phase of cell life cycle, 122–123, *122*
G2 phase of cell life cycle, 122–123, *122*
⁶⁷Ga. See *Gallium, radioactive*
Gallium, radioactive (⁶⁷Ga), radiation dose from, 7t
Gamma rays, from radiopharmaceuticals, 82
  interactions with matter, 10–11
  oxygen enhancement ratio and, 109
  properties of, 11t
  radiotherapeutic uses of, 109
Gastrointestinal syndrome, 134, 136t
  pathology of, 136t, 137–138
  treatment of, 140–143
Gastrointestinal tract, radiation response of, 130, 131–132
  radiosensitivity of, 117–118, 137–138
  radiotolerance of, 151t
Gender. See *Sex*
Gene(s), genic dosage and, 32–34, 38–40
  loss of, 33, *34*
  order on chromosomes, 31–32
    chromosomal aberrations and, 34–36
Gene pool, 26–27
  chromosomal aberrations in, 35–36
  effects of radiation on, 15–30
  genetically significant dose and, 19–23
Genetically significant dose (GSD), 19–23
  determination of, 20–21
  ¹³¹I and, 84
  maximum permissible dose and, 90–92
  mutation frequency and, 21–22, 146
  nuclear medicine and, 83–84
  radiographic examinations and, 21t, 78
Genotype, 17
  of malignant cells, 44
Germ cells, 129
  chromosomal aberrations in, 35
  radiosensitivity of, 25, 27, 143
Glutathione, radioprotection by, 114
Gonadal dose, 2, 25–26, 146
  doubling dose vs., 27, 146
  embryonic dose and, 69
  from radiopharmaceuticals, 7t
  from specific radiographic examinations, 5t
  gene pool and, 18–19
  genetically significant dose and, 20
  nuclear medicine and, 83–84
  radiation-induced mutations and, 26–27
  reduction of, 28
Graves' disease, ¹³¹I treatments, 84–85
Gray (Gy), 3
Growth retardation, radiation-induced, 66–69, 67t, 76
  radiographic examinations and, 78–79
  ultrasound-induced, 157
GSD. See *Genetically significant dose*
Gy, 3

Hair loss, radiation-induced, 134
Health Insurance Plan of New York (HIP) study, 80
Heart, radiotolerance of, 151t
Heat, ultrasound and, 155–156
Hemangiosarcomas, radiation-induced, 59
Hematopoietic system failure, 136t, 137
  pathology of, 134, 136t, 137
  treatment of, 140–143
Hematopoietic tissue, radiation response of, 131, 132
  residual radiation damage to, 152
Hemorrhage, radiation-induced, 136t, 137, *141*
  dose required to produce, 151t
  treatment of, 142–143
Hepatitis, radiation-induced, 151t
Hertz, 155
Heterozygotes, 17–18, *19*
HIP study, 80
Hodgkin's disease, leukemia and, 147–148
Homologous chromosomes, 17, *17*
  meiosis, 35

Homozygotes, 17–18, *19*
Hormones, cancer development and, 45, 53
Hyperbaric oxygenation, 109
Hyperthermia, 114–116, *114*
Hyperthyroidism, treatment with ¹³¹I and, 60
  during pregnancy, 73
  risks of, 84–85
Hypothyroidism, radiation-induced, 84
Hypoxic cells, definition of, 104
  hyperbaric oxygenation and, 109
  in cancers, 106–108, *108*, 113
  radiosensitivity of, 104–111, *105*, *107*, 124
  repair of radiation damage in, 123–124
Hysteria, radiation-induced, 138
Hz, 155

¹³¹I. See *Iodine, radioactive*
ICRP. See *International Commission on Radiation Protection*
Immune system, bone marrow replacement and, 45, 142–143
¹¹³ᵐIn. See *Indium, radioactive*
In utero irradiation. See *Embryo; Embryonic dose; Fetal dose; Fetus*
Indium, radioactive (¹¹³ᵐIn), half-life of, 82
Infection, radiation-induced, 136t, 137–138, *141*
  treatment of, 140–143
Integral dose, 4
International Commission on Radiation Protection (ICRP)
  report, acceptable risk, 89–90
  low-dose radiation, lifetime cancer risk and, 54t
  maximum permissible dose, 89–90, 91t
  risk of radiogenic cancers, 52t
Interphase (G1), 121, *122*
Intestine, large, radiopharmaceutic dose to, 7t
  small, sensitivity to radiation carcinogenesis, 52–53, 52t
Intrauterine death, radiation-induced, 67
Inversion chromosomes, 34–35
Iodine, radioactive (¹³¹I), 6
  diagnostic dose, thyroid cancer and, 59–60
  dilution with nonradioactive isotope, 83
  embryonic dose from, 70t, 84
  fetal thyroid and, 70–73
  genetically significant dose and, 84
  organ distribution of, 6
  placental permeability and, 71
  radiation doses from, 7t
  treatment with, 84–86
Ionizing radiations, matter and energy exchange by, 8–13
  mechanisms of cell killing, 96–99
  types of, 10
  ultrasound vs., 154–156

Kidney, radiation dose to, 7t, 70t
  radiation response of, 132
  radiosensitivity of, 150
  radiotolerance of, 151t
Kinetocores. See *Centromeres*
Klinefelter's syndrome, 40

Late effects of radiation, 134–135, 143–144. See also *Cancer, radiation-induced; Cataracts, radiation-induced; Fertility; Life shortening, radiation-induced*
  on parenchymal or stromal cells, 150
  vascular system and, 149
Latent period for cancer development, 43, 43t, 79
LD₅₀. See *Median lethal dose*
LET. See *Linear energy transfer*
Lethal dose, median. See *Median lethal dose*
Lethal radiation damage, 94–102
  LET and, 103–104, *104*
  quasi-threshold radiation dose and, 100, *100*, 124–125
  single hit killing and, 94
  to organs and tissues, 129
Leukemia, chronic myeloid, 44
  Hodgkin's disease and, 147–148
  radiation-induced, 49, 51
    ankylosing spondylitis treatment and, 55t, 56
    atomic bomb survivors and, 53, 56
    diagnostic x rays and, 58–59
    ¹³¹I treatments and, 60, 85
    in utero x irradiation and, 55t, 58

Leukemia, radiation-induced *(continued)*
  latent period for, 43, 43t
  nuclear medicine and, 51, 83–84
  occupational radiation exposure and, 91–92
  ^{32}P treatments and, 60, 86
  radiographic examinations and, 78
  radiologists and, 60–61
  risk for, 52t, 55t
  thorium and, 59
Leukopenia, radiation-induced, 136t
Life shortening, radiation-induced, 91t, 92, *93*
  total-body radiation and, 134–135, 144
Life Span Study of atomic bomb survivors, 55
Linear energy transfer (LET), 12–13
  biologic damage and, 12
  chromosome aberrations and, 36–38
  dose rate and, 116–117
  dose-survival relationships and, 94–99, 103–104
  mutation frequency and, 24–25
  oxygen enhancement ratio and, 105–106, *107*
  radiation damage and, 103–104, *104*
  radiotherapy and, 109–111
  values of ionizing particles, 13t
  variation with RBE, *14*
Liver, concentration of ^{99m}Tc, 6
  radiation response of, 132
  radiotolerance of, 151t
  sensitivity to radiation-induced cancers, 52–53, 52t
Liver dose, from radiopharmaceuticals, 7t, 86
Low-dose irradiation. See *Nuclear medicine; Radiographic examinations*
Lung, radiation dose to, from radiopharmaceuticals, 7t
  radiotolerance of, 151t
  sensitivity to radiation carcinogenesis, 52–53, 52t
Lung cancer, radiation-induced, atomic bomb survivors and, 56
  risk in particular populations, 55t, 61
Lung scans, radiation-induced leukemia and, 83–84
Lymphatic tissue, differentiation in, 130
  sensitivity to radiation carcinogenesis, 52–53, 52t
Lymphocytes, total-body radiation and, 140, *141*
Lymphoma, radiation-induced, risk in particular populations, 55t

Malformations, congenital. See *Congenital malformations*
Malignant cells, chromosome aberrations in, 39–40
Malignant transformation, 126–127
Mammography, 79–81
  radiation dose from, 5t
Mammotrophic hormone, in cancer development, 45
Marshall Islanders, radiation-induced cancers in, 55t, 56, 60
Mastitis, radiation therapy and, 55t, 56, 80t
Maximum permissible dose (MPD), 89–93
  acceptable risk and, 89
  ALARA concept, 91
  cancer induction and, 91–92
  cataract induction and, 92–93
  fetal, 91
  genetically significant dose and, 90–92
  guidelines for, 89–90, 90, 91t
  life shortening and, 92
  sterility and, 93
  stochastic vs. nonstochastic effects, 89
Median lethal dose (LD$_{50}$), 135, *135*
Meiosis, chromosome aberrations and, 34–36
Meningitis, radiation-induced, 136t
Mental retardation, radiation-induced, 68
2-Mercaptopropionic (MPG), radioprotection by, 114
Metronidazole, 112–113
Microcephaly, radiation-induced, 68
Micronucleus, 38
Microphthalmia, radiation-induced, 68
Microwaves, heat induction by, 115
Midline dose, 4
Midplane dose, 4
Misonidazole, 112–113
Mitosis, *39*
  cell life cycle and, 122, *122*
  chromosome aberrations and, 34–36
  density inhibition and, 126–127
  injured chromosomes in, 38–40
  radiosensitivity of cells during, 122–123, *122*, 130
Mitotic death, 33, 131–132, 151

MPD. See *Maximum permissible dose*
Muscle, radiation response of, 130, 132
Mutagens, carcinogens vs., 44
Mutations, 15–40. See also *Chromosome aberrations; Genetically significant dose*
  dominant, 17, 28t
    gonadal dose and, 26–27
  elimination from gene pool, 25, 35–36
    rate of, 16–19
  per rem ionizing radiation, 22–24
  point, 15–19, 25–26
  premutational changes, 24–25
  recessive, 17, 28t
  repair of, 24–25, 78
    Q theory, 101
  spontaneous, 15–16, 23
  X-linked, 26
Mutation frequency, 22–26, 147
  changes in, 16–19
  dose rate and, 24–25
  genetically significant dose and, 21–22, 146
  LET and, 24–25
  radiation dose and, 22
Mutation theory of malignant growth, 44

National Council on Radiation Protection and Measurements (NCRP), maximum permissible dose, 90, 90t
Nausea, radiation-induced, 135
NCRP. See *National Council on Radiation Protection and Measurements report*
Negative pi-mesons, 11t, 111
Negatrons, properties of, 11t
Neonatal death, radiation-induced, 66–67, *66*
Nephrosclerosis, radiation-induced, 151t
Neutrons, 8, 11t
  cataract induction and, 92
  fast vs. slow, 11
  LET value of, 13t
  oxygen enhancement ratio and, 109
  properties of, 11t
  radiotherapeutic uses of, 109–110
Neutrophils, total-body radiation and, 140, *141*
Nondisjunction of chromosomes, 38–40, *39*
Nonparticulate radiations, 10
Nonstochastic radiation effects, 89
Nuclear medicine. See also *Radiopharmaceuticals*
  genetically significant dose and, 83–84
  risks of, benefits vs., 75–76
    cancer induction by, 83–84
    diagnostic procedures and, 82–84
    reduction of, 82–83
Nuclei, heavy, properties of, 11t

Occupational radiation exposure, 51. See also *Maximum permissible dose*
  cancer induction by, 55t
  chromosome aberrations and, 38–40
  risks of, 89–93, 91t
OER. See *Oxygen enhancement ratio*
Oncogenic viruses, 44
Optical orbits, 9
Organ-of-interest dose, 4
Organs. See also names of specific organs
  radiation response of, 132–136
    late effects, 149
    late reactions, 133
Osteosarcomas. See *Bone cancer*
Ovarian cancers, radiation-induced, *49*
Ovaries, irradiation for castration, 57
Oxygen. See also *Oxygen enhancement ratio*
  chromosome aberrations and, 39–40
  dose-modifying effect of, 104–111, *105*
  hyperbaric oxygenation and, 109
  supply to cancer cells, 106–108, *108*
Oxygen enhancement ratio (OER), calculation of, 105
  cell life cycle and, 123
  for gamma rays, 109
  for negative pi-mesons, 111
  for neutrons, 109
  for x rays, 109

LET dependence and, 105–106, *107*

^{32}P. See *Phosphorus, radioactive*
Pancarditis, radiation-induced, 151t
Pancytopenia, radiation-induced, 151t
Parenchymal tissue, radiosensitivity of, 132, 149–150
Parental radiation, 26–27, 28t
Particulate radiation, 10–12
Pelvic cancers, radiotherapy of, 111
Pericarditis, radiation-induced, 151t
Perinatal death, radiation-induced, 66, 76
Phenotype, 17
Philadelphia chromosome, 44
Phosphorus, radioactive (^{32}P), cancer induction by, 55t, 60
    treatment of polycythemia vera with, 60, 86
Pituitary cancers, radiotherapy of, 110
Placenta, permeability to ^{131}I, 71
Platelets, total-body radiation and, 140, *141*
Pneumonitis, radiation-induced, dose required, 151t
Polycythemia vera, leukemia and, 86
    ^{32}P treatment of, 60, 86
Polysomy, 38
Positrons, properties of, 11t
Post-irradiation conditions, density inhibition and, 126–127
    repair of radiation damage in, 120–121
Potentially lethal radiation damage, 119–120
    density inhibition and, 126–127
    malignant transformation and, 126–127
    repair of, 119–120
        in hypoxic cells, 124
    residual radiation damage vs., 151
    sublethal radiation damage vs., 119–120, 125
    targets of, 125
Pregnancy. See also *Embryo; Fetus*
    ^{131}I treatment during, 84
    radiation therapy during, 148–149
        benefits vs. risks, 76
    radiographic examinations during, 71–72, 78–79
    therapeutic abortions after radiation, 72–73
    ultrasound exposure during, 157
Project Smoky, 51
Proliferating cells, density inhibition and, 126–127
    in cancers, *108*
    in organs and tissues, 129–133
    radiosensitivity and, 131
    repair of radiation damage in, 120, 151
Promoters, development of latent tumor cells into cancer and, 45
Prostate, radiosensitivity of, 150
Protons, 8
    LET value of, 13t
    properties of, 11t
    radiotherapeutic uses of, 111
Pulmonary embolism, risks of lung scan with, 83–84

Q theory of radiation damage, 101, 123
    LET and, 103–104
QF, 3
Quality factor (QF), 3
Quasi-threshold dose, 100, 124–125, *126*
Quiescent cells. See also *Differentiated cells*
    in cancers, *108*
    in organs and tissues, radiation response of, 129–133
    repair of radiation damage in, 120, 151

Ra. See *Radium*
Rad, 3
Radiation damage. See also *Lethal radiation damage; Potentially lethal radiation damage; Residual radiation damage; Sublethal radiation damage*
    LET and, 103–104, *104*
    repair of, post-irradiation conditions and, 120
Radiation dose, 2–3. See also *Dose-survival relationships;* names of specific elements
    from radiopharmaceuticals, 5–6, 7t, 82
    in diagnostic roentgenology, 3–5, 77
    maximum permissible, 89–93
    mutation frequency and, 22–26
    population exposure estimates, 1t

rate. See *Dose rate*
Radiation response. See also names of specific organs and tissues
    factors influencing, 2. See also *Age; Sex*
Radioactive decay, radiation produced by, 11t
Radiodermatitis, 60
Radiofrequency heat induction, 115
Radiographic examinations, cancer induction by, 55t, 58–59, 78
    estimated radiation doses of, 5t
        embryonic dose from, 70t
    genetically significant dose and, 21t, 78
    pregnancy and, 71–73, 78–79
    risks of, 75–79, 77
        reduction of, 81–82
    screening vs. diagnostic use, 79–82
    yearly total in United States population, 76–77
Radiologists, occupational radiation exposure of, cancers associated with, 55t
    leukemia and, 60–61
    life shortening and, 92, *93*, 144
    radiodermatitis and, 60
    skin cancer and, 60
Radionuclides. See *Radiopharmaceuticals*
Radiopharmaceuticals. See also names of specific elements
    radiation dose from, 5–6, 7t
        embryonic dose of, 70, 70t
    risk of, factors affecting, 82
        reduction of, 82–84
    yearly individual exposure to, 1t
Radioprotectors, 114
    anesthetics as, 109
    sulphydryl compounds as, 114, 123
Radiosensitivity, 99–100. See also names of specific organs
    calculation from dose-survival curves, *99*
    cell life cycle and, 121–123, *122*
    oxygen effect on, 104–111, *105*
Radiosensitizers, 111–114
Radiotherapy. See also names of specific elements
    charged particle therapy, 110–111
    dose-rate effect and, 116–118
    fractionated, 108–109, 113
    high-LET radiations in, 109–111
    hyperbaric oxygenation and, 109
    hyperthermia and, 114–116
    neutron therapy, 109–110
    radioprotectors and, 114
    radiosensitizers and, 111–114
    risks of, 146–152, *147–148*
Radium, injections of, cancer induction and, 55t, 59
    pelvic implants of, 57
Radium dial painters, occupational radiation exposure, 55t, 61
RBE. See *Relative biologic effectiveness*
Rectum, sensitivity to radiation carcinogenesis, 52–53, 52t
Relative biologic effectiveness (RBE), 3, 13, *14*
Relative mutation risk, 23
Rem, 3
Residual radiation damage, 150–152
Restituted chromosomes, 33
    dose-rate effect and, 36–37
Ring chromosomes, 34–35
Ringworm, x ray treatment of, 57
Roentgen, definition of, 2–3
"Runt" disease, 142

S phase of cell life cycle, 122, *122*
    radiosensitivity of cells and, 122–123, *122*
^{75}Se. See *Selenomethionine*
Selenomethionine (^{75}Se), radiation dose from, 7t
Sex. See also *Pregnancy*
    influence on radiation response, 2, 25
        carcinogenesis and, 53
Sex ratio, parental radiation and, 26
SI, 3
Sievert, 3
Skin, radiation response of, 132
    radiosensitivity of, 151
Skin cancer, radiation-induced, 55t, 60
Skin dose, 4, 5t
Soft tissue, x ray absorption of, 3
Somatic cells, chromosome aberrations in, 35
Somatic mutations, carcinogenesis and, 43
Spermatogenesis, radiosensitivity and, 25, 143

Spinal cord, radiotolerance of, 151t
Spindle fibers, 34, 38–40
Spleen dose, from ^{51}Cr-sodium chromate, 7t
  ^{32}P treatments and, 86
Stem cells, 129–132
Sterility, radiation-induced, 143, 146. See also *Fertility*
  in utero irradiation and, 69
  late effect of total-body radiation, 134–135
  maximum permissible dose and, 93
Stochastic radiation effects, 89
Stromal cells, radiosensitivity of, 150
Sublethal radiation damage, 95–99, 119–121
  cellular targets and, 97–101
  LET and, 103–104, *104*
  maximum, 97, 100, 120
  measurement of, 124–125
  potentially lethal radiation damage vs., 119–120, 125
  quasi-threshold radiation dose and, 100, *100*, 124–125
  repair of, 116–124, *120*
  residual radiation damage vs., 150–151
  targets of, 125
Sulphydryl compounds, radioprotection by, 114, 123
Synapsis, 35
Systeme Internationale (SI), 3

Target theory of radiation damage, 97–101
  LET and, 103–104
^{99m}Tc. See *Technetium, radioactive*
^{201}Te. See *Tellurium, radioactive*
Technetium, radioactive (^{99m}Tc), embryonic dose of, 70t
  radiation dose from, 7t
  concentration in body organs, 83
  half-life of, 82
  organ distribution of, 6
  radiation emitted by, 82
Tellurium, radioactive (^{201}Te), chloride, radiation dose from, 7t
Testicular atrophy, radiation-induced, 68
Th. See *Thorium*
Thermotolerance of cells, 114–116, *114*
Thorium, as contrast agent, cancer induction by, 55t, 59
  treatment with, cancer induction by, 55t
Thorotrast. See *Thorium*
Threshold dose, 89
Thrombocytopenia, radiation-induced, 136t
Thymus, enlarged, x-irradiation therapy for, 57
Thyroid, adenocarcinomas of, 52–53, 52t
  adenomas of, radiation-induced, 57
  concentration of ^{131}I in, 6
  fetal, sensitivity to ^{131}I, 71, 73
  radiation dose to, from radiopharmaceuticals, 7t
  radiosensitivity of, 57
  x-irradiation of, 55t, 57
Thyroid cancer, radiation-induced, 57
  atomic bomb survivors and, 56
  ^{131}I treatments and, diagnostic dose, 59–60
    for hyperthyroidism, 60, 85
  latent period of, 43, 43t
  prenatal ^{131}I exposure and, 71
  risk in particular populations, 55t
  spontaneous, treatment with ^{131}I, 60, 85–86
Thyroid stimulating hormone, in cancer development, 45
Thyroiditis, radiation-induced, 84
Tinea capitis, x-ray treatment of, 57
Tissues. See also names of specific tissues
  absorption of ultrasound, 156
  radiation response of, 129–133
    effect of tissue volume on, 2
  radiotolerance of, 151t
  residual radiation damage to, 150–151

Tonsils, enlarged, x-irradiation therapy for, 57
Total-body dose, nuclear medicine and, 83–84
Total-body radiation. See also *Acute radiation syndromes*
  acute effects of, 134–143, *141*, 149
    indicators of radiation dose, 140–141
    treatment of, 139–140
  cancer induction by, 143
  cataract induction by, 134–135, 144
  fertility impairment by, 143
  influence of dose on mortality, 135
  late effects of, 134–135, 143–144, 149
    vascular damage, 149
  latent period for acute radiation syndromes, 135–136, *136*
  medium lethal dose and, 135, *135*
  prodromal reaction to, 135, *136*
  survival possibilities and, 139–140
Toxic nodular goiter, 6
Translocation chromosomes, 34–35
Tuberculosis, radiation-induced cancers and, 55t, 56

Ultrasound, 155–156
  heat induction by, 115, 155–156
  ionizing radiations vs., 154–156
  pregnancy and, 157
  risks of, 154–157
UN Scientific Committee on the Effects of Atomic Radiation
  (UNSCEAR) report, doubling dose, 24
  effect of fetal irradiation, 58
  low-dose radiation, 52
    lifetime cancer risk, 54t
  radiation-induced mutations, 28t
  reduction of genetic risks, 28
  risk of radiogenic cancers, 52t
United States population, annual bone marrow dose to, 77
  yearly radiation exposure of, 1, 1t
    by diagnostic x rays, 76–77
UNSCEAR report. See *UN Scientific Committee on the Effects of Atomic Radiation report*
Uranium miners, occupational radiation exposure in, 55t, 61

Van de Graaff generators, radiation produced by, 11t
Vascular system, in cancers, 106–108, *108*
  radiosensitivity of, 138, 149
Vasculitis, radiation-induced, 136t
Viruses, oncogenic, release by radiation, 44
Vomiting, radiation-induced, 135

Water, radiation-produced free radicals, 10
Weight loss, radiation-induced, 134

X rays. See also *Radiographic examinations*
  absorption by body tissues, 3–5, *4*
  interactions with matter, 10–11
  oxygen enhancement ratio and, 109
  properties of, 11t
  radiotherapeutic uses of, 109
  yearly individual exposure to, 1t
X-linked mutations, 26
  in cancer cells, 44
^{133}Xe. See *Xenon, radioactive*
Xenon, radioactive (^{133}Xe), radiation dose from, 7t

Z value of an atom, 8–9